CLIMATE CHANGE, ADAPTIVE CAPACITY AND DEVELOPMENT

Sponsors:

Canadian International Development Agency
Climate Change Unit, Environment Division, World Bank
Electric Power Research Institute, USA
Federal Ministry for Economic Cooperation and Development/
Gesellschaft für Technische Zusammenarbeit, Germany
Organisation for Economic Cooperation and Development
United Nations Development Programme

Co-sponsors:

Adaptation and Impacts Research Group, Environment Canada
Neitherlands Climate Change Studies Assistance Programme
United Nations Environment Programme

CLIMATE CHANGE, ADAPTIVE CAPACITY AND DEVELOPMENT

editors

Joel B Smith
Stratus Consulting Inc, USA

Richard J T Klein
Potsdam Institute for Climate Impact Research, Germany

Saleemul Huq
International Institute for Environment and Development, UK

ICP

Imperial College Press

Published by

Imperial College Press
57 Shelton Street
Covent Garden
London WC2H 9HE

Distributed by

World Scientific Publishing Co. Pte. Ltd.
5 Toh Tuck Link, Singapore 596224
USA office: Suite 202, 1060 Main Street, River Edge, NJ 07661
UK office: 57 Shelton Street, Covent Garden, London WC2H 9HE

British Library Cataloguing-in-Publication Data
A catalogue record for this book is available from the British Library.

CLIMATE CHANGE, ADAPTIVE CAPACITY AND DEVELOPMENT

ISBN 1-86094-373-X

Printed in Singapore by World Scientific Printers (S) Pte Ltd

Contents

Acknowledgments

This book reflects the hard work of not just the authors and editors but also many other people. The book would not have been possible without the workshop, "Enhancing the Capacity of Developing Countries to Adapt to Climate Change", held in Potsdam, Germany, from September 30 to October 2, 2001. The workshop created an intellectual environment to inspire the chapters and the discussion reflected in the summary chapter. That workshop was followed with a meeting of the authors in London in March 2002 to discuss reviewer comments and further refine the summary chapter.

The workshop was made possible by the tireless work and attention to detail of the Potsdam Institute for Climate Impact Research (PIK). We wish to especially thank Prof. Dr. Hans Joachim Schellnhuber, Director of PIK (now on leave at the Tyndall Center). John created the first "Potsdam Sustainability Days", which included this adaptation workshop, and PIK staff handled the preparations and logistics for the workshop. We also acknowledge the work of Ferenc Toth (now at the International Institute for Applied Systems Analysis) in conceptualizing and organizing the workshop. We also gratefully acknowledge the work of Johann Grüneweg from PIK in handling the many travel, hotel and other logistical arrangements for the Potsdam workshop.

The authors' meeting in March 2002 was organized by the International Institute for Environment and Development (IIED) in London. We gratefully acknowledge the contributions of the staff in handling hotel and meeting site arrangements. In particular, we wish to mention Sue Mylde's assistance in coordinating hotel and meeting arrangements.

All of the papers were peer reviewed, either by other authors in this book or by external peer reviewers. We wish to thank the following external reviewers for their thoughtful and constructive comments: Stewart Cohen, Sally Kane, Robert Kates, Sam Fankhauser, Neil Leary, Liza Leclerc, Elizabeth Malone and Richard Tol.

For the book, we want to particularly acknowledge the hard work and perseverance of the book's technical editor, Christina Thomas of Stratus

Consulting. Christina provided a consistent and steady hand throughout the process and devoted many hours to ensuring the quality of the book. Christina was assisted by Stratus Consulting's Document Solutions staff of Diane Blagusz, Sara Garland and Erin Miles. Shiela DeMars' assistance with organizing the Potsdam workshop and the authors' meeting as well as handling administrative matters throughout the project was invaluable.

Finally, we wish to acknowledge the generous financial contributions and patience of our sponsors and cosponsors. Without their support and confidence, we would not have been able to start or finish this effort.

Joel B. Smith
Richard J.T. Klein
Saleemul Huq

1. Introduction

Joel B. Smith
Stratus Consulting
Boulder, Colorado, USA

Richard J.T. Klein
Potsdam Institute for Climate Impact Research
Potsdam, Germany

Saleemul Huq
International Institute for Environment and Development
London, United Kingdom

The increased concentrations of greenhouse gases in the atmosphere threaten to dramatically change the earth's climate in the 21st century. Indeed the climate of the earth already changed in the 20th century, with half of the most recent warming caused by increased greenhouse gas concentrations (Houghton et al., 2001). The Intergovernmental Panel on Climate Change (IPCC) projects that warming in this century could range from 1.4 to 5.8°C (Houghton et al., 2001). Accompanying the warming will be substantial increases in sea level and average precipitation and also changes in climate variability, storms, floods, cyclones, etc. Further climate change is virtually inevitable. The rise in greenhouse gas concentrations to date commits the world to approximately another 1°C of rise in mean global temperature (Hansen et al., 2002). Even the most feasible aggressive emission control measures will still allow for a further rise in greenhouse gas concentrations and additional warming and sea level rise (Metz et al., 2001).

With the climate already changing and further change in climate highly likely to happen, adaptation is a necessary component of any response to climate change. However, mitigation is needed to avoid the worst effects of climate change.

Many countries and specific sectors within these countries are vulnerable to the effects of climate change. Vulnerability is a function of (McCarthy et al., 2001):

- *Exposure*. Exposure is the change in climate and what is affected, e.g., how many people are living in an area that could be inundated by sea level rise or affected by change in coastal storms.
- *Sensitivity*. Sensitivity is the direct effect of climate change on systems, e.g., changes in crop yields or runoff.
- *Adaptive capacity*. Adaptive capacity is the ability of a system to adapt to climate change, reduce adverse effects or take advantage of beneficial effects.

The IPCC (McCarthy et al., 2001) concluded that developing countries are much more vulnerable to climate change. Some of this vulnerability has to do with greater exposure to climate change, e.g., having large populations in low lying coastal areas exposed to sea level rise. Some of it has to do with greater sensitivity, e.g., having crops closer to thermal thresholds beyond which yields decline. It is fair to say that most of the relatively greater vulnerability of developing countries is a function of their more limited adaptive capacity.

The IPCC (Smit et al., 2001, p. 879) also states,

> Adaptation to climate change has the potential to substantially reduce many of the adverse impacts of climate change and enhance beneficial effects—though neither without cost nor without leaving residual damage. . . . Adaptation depends greatly on the *adaptive capacity* or adaptability of an affected system, region or community to cope with the impacts and risks of climate change. The adaptive capacity of communities is determined by their socioeconomic characteristics. Enhancement of adaptive capacity represents a practical means of coping with change and uncertainties in climate, including variability and extremes. In this way, enhancement of adaptive capacity reduces vulnerabilities and promotes sustainable development. . . . Enhancement of adaptive capacity is a necessary condition for reducing vulnerability, particularly for the most vulnerable regions, nations and socioeconomic groups.

The IPCC further states that adaptive capacity is a function of:

- wealth
- access to technology
- stable and effective institutions
- systems in place for dissemination of information
- equitable distribution of power
- well-functioning social systems (e.g., Adger, 1999).

Developing countries generally possess less of these attributes of adaptive capacity than do developed countries. In particular,

- Developing countries are poorer than developed countries. For example, the countries belonging to the Organisation for Economic Cooperation and Development, generally the most developed countries, contain 1.1 billion people, or 19 percent of global population, but have 81 percent of global product (WRI, 2000). The remaining countries, which account for over four-fifths of the world's population and include all of the developing countries, account for less than one-fifth of the world's income.
- Developing countries generally have less access to technology.
- Institutions in developing countries tend to be less stable (although democracy has spread to many countries). Many have limited or no democracy.
- Information dissemination systems in developing countries are less advanced.
- Finally, developing countries have less equality in income distribution[1] (WRI, 2000).

Clearly, based on these indicators of adaptive capacity, developing countries have much less of it than developed countries.

The question arises of how best to help developing countries adapt to climate change. Should the focus be on specific adaptations to climate change? This would mean investing in such measures as planning for sea level rise, enhancing evacuation procedures, developing heat or drought resistant crops and enhancing robustness of water supply systems so that the capability to cope with specific impacts on climate sensitive sectors is enhanced. Reading Chapter 18 of the IPCC Working Group II (Smit et al., 2001), one might ask, How successful would such specific adaptations be if adaptive capacity is limited? Can specific adaptations be successfully implemented, maintained and made permanent if income, access to technology and information and stability of institutions are limited?

The alternative is to invest in enhancing adaptive capacity, e.g., increasing wealth and improving education, income distribution, institutions and health care so that society is better prepared to cope with climate change and other stresses. These can be thought of as investments in development. Would such investments enhance adaptive capacity and reduce vulnerability of developing countries?

Ultimately, the question becomes whether it is more prudent to invest in specific adaptations or enhancements of adaptive capacity. These questions are quite relevant because the Conference of the Parties to the United Nations Framework Convention on Climate Change (UNFCCC) has set up several funds to support adaptation to climate change (www.unfccc.int). How should those funds be most effectively disbursed to support adaptation to climate change? How can those funds most effectively reduce vulnerability?

This is an emerging topic and one on which there is little research. A critical need is to improve our understanding of how best to help developing countries adapt to climate change and how to best enhance their adaptive capacity.

To address these questions, Stratus Consulting and the Potsdam Institute for Climate Impact Research (PIK) organized a workshop of leading researchers on adaptation and representatives from developing countries, donor countries, multilateral agencies and nongovernmental organizations. The goal of the workshop was to develop a research agenda on the question of how best to enhance the adaptive capacity of developing countries.

The workshop (and a subsequent meeting of authors) was funded through the generous sponsorship of the following organizations:

- Canadian International Development Agency
- Climate Change Unit, Environment Division, World Bank
- Electric Power Research Institute, USA
- Federal Ministry for Economic Co-operation and Development/ Gesellschaft für Technische Zusammenarbeit, Germany
- Organisation for Economic Cooperation and Development
- United Nations Development Programme.

In addition, the following organizations cosponsored the workshop:

- Adaptation and Impacts Research Group, Environment Canada
- Netherlands Climate Change Studies Assistance Programme
- United Nations Environment Programme.

The workshop was held on PIK's campus in the Telegrafenberg section of Potsdam, Germany, from September 30 to October 2, 2001. Forty-five individuals (including two by telephone) attended the workshop. An agenda and list of participants are included in the appendix.

A number of topics were addressed at the workshop and in commissioned papers, including the following:

- Is vulnerability to climate change reduced more by investing in specific adaptations or in broader development?
- To what degree do differences in risks of climate change, nonclimate stresses and levels of development lead to differences in adaptation requirements across countries and regions?
- How can adaptation assistance be tailored to be appropriate for local needs?

The workshop began with overview papers on analytic and methodological issues. These papers addressed such topics as social aspects of adaptive capacity. Following that, papers on links with nonclimate issues were presented. These papers covered such topics as lessons from natural hazards management and famine early warning and food security. Then, a series of case studies addressed country and even community specific vulnerabilities to climate variability and change in the context of adaptive capacity. The workshop concluded with a series of panel discussions on research needs. There were three panels, consisting first of researchers, then representatives from nongovernmental organizations, then representatives from multilaterals and donor institutions.

The 13 chapters in this book are based on papers presented at the workshop. The first set of chapters addresses analytic and methodological issues. These give an overview of issues to be considered in addressing enhancement of adaptive capacity. Chapter 2, by Barry Smit and Olga Pilisofova, reviews Chapter 18 of the IPCC Working Group II report (Smit et al., 2001) and lessons it gives for the questions raised in the workshop. Chapter 3, by W. Neil Adger, examines the role of social capital in adaptation to climate variability and climate change.

A second set of chapters addresses links with nonclimate issues, drawing lessons from sectors already addressing similar risks. Chapter 4, by John Handmer, concerns the lessons that management of natural hazards offers to adaptation to climate change. Indeed, much of adaptation to climate change will be adapting to changes in extreme climate events that the natural hazards management community is already addressing (e.g., Mileti, 1999). Chapter 5, by Thomas E. Downing, focuses on food security and how conventional assessment approaches have not identified vulnerability at different scales. Appropriate adaptations can differ, depending on whether national, regional, individual or household levels are being considered. Chapter 6, by Gary Yohe, Kenneth Strzepek, Tammy Pau and Courtney Yohe, quantitatively analyzes how potential future changes in socioeconomic conditions can affect adaptive capacity, considering Egypt. Chapter 7, by Ian Burton, examines the ability of countries to apply their adaptive capacity. Possessing adaptive capacity is no guarantee that it will be used wisely. Chapters 8 and 9, by Roda Verheyen, explore the legal framework in the UNFCCC and subsequent agreements for support by developed countries of adaptation activities and funding in developing countries.

The third set of chapters presents case studies from developing countries. They give a flavor of the issues and challenges in enhancing adaptive capacity as well the differences in physical and human elements that are present across the world. Chapter 10, by Rosa T. Perez, examines the relative roles of the national and regional governments in addressing adaptation to climate variability and climate change in the coastal resources of the Philippines. Chapter 11, by Cecilia Conde and Hallie Eakin, focuses on how recent socioeconomic trends, such as the North American Free Trade Agreement, have affected the capacity of farmers in one Mexican state to cope with variability from the El Niño/Southern Oscillation and with climate change. Chapter 12, by Chris H. D. Magadza, explores how the government and political environment in Africa can affect the ability of countries to address climate change impacts and adaptation. Chapter 13, by Shaheen R. Khan, addresses how political developments in Pakistan affect its capacity to implement specific natural resources management measures to reduce risks from climate variability and climate change.

The final chapter contains a summary of the discussion and recommendations from the workshop.

Endnotes

[1] This is determined by examining the Gini coefficient, a measure of the distribution of wealth. Developing countries typically have less equitable distributions of wealth. However, there is a wide distribution of Gini coefficients and a number of developing countries have more equitable distributions of wealth than a number of developed countries (WRI, 2000).

References

Adger, W. N. (1999), Social vulnerability to climate change and extremes in coastal Vietnam. *World Development* **27**, 249-269.

Hansen, J. (and 27 co-authors) (2002), Climate forcings in GISS SI2000 simulations. *Journal of Geophysical Research—Atmospheres* **107**, 4347-4384.

Houghton, J. T., Ding, Y., Griggs, D. J., Noguer, M., van der Linden, P. J., Xiaosu, D. and Maskell, K., eds. (2001), *Climate Change 2001: The Scientific Basis*. Cambridge University Press, New York.

McCarthy, J., Canziana, O., Leary, N., Dokken, D. and White, K., eds. (2001), *Climate Change 2001: Impacts, Adaptation, and Vulnerability*. Cambridge University Press, Cambridge, United Kingdom.

Metz, B., Davidson, O., Swart, R. and Pan, J. (2001), *Climate Change 2001: Mitigation*. Cambridge University Press, Cambridge, United Kingdom.

Mileti, D. (1999), *Disasters by Design: A Reassessment of Natural Hazards in the United States*. Joseph Henry Press, Washington, DC.

Smit, B., Pilifosova, O., Burton, I., Challenger, B., Huq, S., Klein, R. and Yohe, G. (2001), Adaptation to climate change in the context of sustainable development and equity. In *Climate Change 2001: Impacts, Adaptation, and Vulnerability*, McCarthy, J., Canziana, O., Leary, N., Dokken, D. and K. White, eds. Cambridge University Press, Cambridge, United Kingdom. 877-912.

WRI (2000), *World Resources: 2000-2001*. World Resources Institute, Washington, DC.

2. From Adaptation to Adaptive Capacity and Vulnerability Reduction

Barry Smit
University of Guelph
Guelph, Ontario, Canada
Olga Pilifosova
United Nations Framework Convention on Climate Change
Bonn, Germany

Adaptation to climate change includes adjustments in socioeconomic systems to reduce their vulnerability both to long-term shifts in average climate and to changes in the frequency and magnitude of climatic extremes. In many parts of the world, especially in developing countries, these extremes are hazardous now, and often exceed the capacity of a country or community to cope. The vulnerability of a community to climate change is related to the exposure of the community to hazardous climatic conditions and to the adaptive capacity of the community to deal with those conditions. Enhancing the ability of communities to adapt to climate change or manage climate change risks requires addressing pertinent locally identified vulnerabilities, to involve stakeholders and to ensure that adaptation initiatives are compatible with existing decision processes.

1. Introduction

Adaptation to climate change refers to adjustments in ecological, social and economic systems in response to actual or expected climatic stimuli and their effects or impacts. Adaptation is important in the climate change issue in two ways (Figure 1): one deals with autonomous adaptation as part of impact and vulnerability assessment, and the other deals with adaptation as a planned policy response. Adaptations are widely undertaken autonomously, without consciously planned interventions, and these are often expected to have

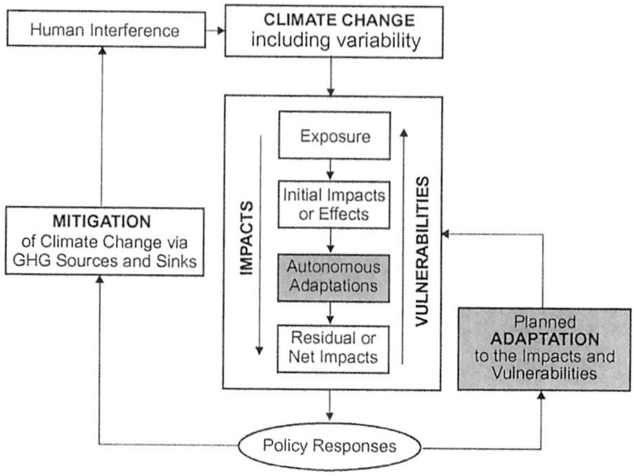

Figure 1. Adaptation in the climate change issue (based on Smit et al., 1999; Smit and Pilifosova, 2001)

potential to reduce negative climate change impacts or to moderate vulnerabilities, thereby easing concerns over the danger of climate change. However, research summarized in the Intergovernmental Panel on Climate Change (IPCC) Third Assessment Report (Smit and Pilifosova, 2001) indicates that autonomous adaptation tends to be reactive, can be quite costly and would usually incur some residual damages.

The other place of adaptation in the climate change issue is as a consciously planned response, either reactive or anticipatory, to expected impacts or vulnerabilities of climate change (Figure 1). In negotiations about climate change, the two broad policy responses debated are mitigation (of greenhouse gas emissions) and adaptation (to climate change impacts or risks). Countries are committed in the United Nations Framework Convention on Climate Change (UNFCCC) to formulate and implement measures to facilitate adaptation (Article 4.1b), and to cooperate in preparing for adaptation to the impacts of climate change (Article 4.1e). Policy initiatives to facilitate or promote adaptation are being considered at local, national and international scales.

Several interrelated developments in the field of climate change adaptation are evident over the past several years, especially as lessons have been gained from social and economic risk management (Smit and Pilifosova, 2001). One development is the recognition that for the purposes of adapting to climate change risks in most countries and communities, the key adaptations are less often those related to changes in long-term average temperature and more often related to the frequency and magnitude of extremes such as droughts or floods. Developing countries, in particular, are already vulnerable to these extremes, many of which are likely to become more frequent and more severe with climate change.

A second, related development has been in the approach to climate change impact assessment. Many insights have been gained from impact assessments that start with climate change scenarios (usually average temperature), estimate biophysical effects, then predict the economic impacts of these, and then postulate particular adaptation measures to address negative impacts. A complementary approach starts by identifying vulnerabilities of the system or community to climate, considers current adaptive strategies and their prospects for managing changes in vulnerabilities, then links adaptive management options to decision processes already in place. This vulnerability approach is proving helpful to those who wish to undertake adaptation initiatives that address locally relevant climate-related hazards.

A third development has been a move from considering particular adaptation measures or practices, which can be almost infinite in number, to the assessment and promotion of adaptive capacity. Adaptive capacity is the potential or ability of a system, region or community to adapt. Enhancement of adaptive capacity represents a practical means of coping with changes and uncertainties in climate, including variability and extremes.

In this paper we describe these interrelated developments, and in the process present more formal treatments of the concepts of vulnerability and adaptive capacity. We conclude by showing the ways in which these concepts relate to capacity enhancement and to economic development.

2. Climate Stimuli for Adaptation

Conventionally, impact and adaptation studies have been based on climate change scenarios that provide a limited set of possible future climates—usually specified as average annual conditions such as temperature and moisture. Yet the climate change related stimuli for which adaptations are undertaken are not limited to changes in average annual conditions; they include variability, particularly interannual variability, and associated extremes (Chapters 5 and 12 of this volume). Climatic conditions are inherently variable, from year to year and decade to decade. Variability goes along with and is an integral part of climate change (Hulme et al., 1999). A change in mean conditions is actually experienced through changes in the frequency and magnitude of conditions each year, including extremes (Figure 2).

Figure 2 illustrates the interannual variation in a community-relevant climate attribute (*X*) such as drought severity for a hypothetical system (a community in a region) over approximately 80 years. In this case, some years are drier than others and some are wetter. The average drought severity value rarely occurs. The time-series shows a gradual increasing trend in drought severity, such that the future mean ($\overline{X}_f$) is higher than the present mean ($\overline{X}_p$).

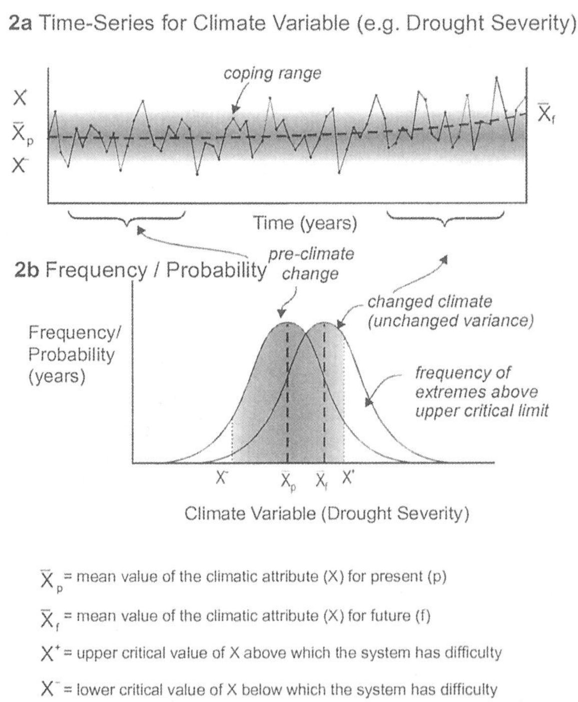

Figure 2. Climate change, extremes and coping range (based on Hewitt and Burton, 1971; Fukui, 1979; Smit et al., 2000; Smit and Pilifosova, 2001)

This system, like most communities and activities, can cope with conditions that deviate from average, but only to some degree. The system has evolved over the years so that somewhat drier and somewhat wetter years are not greatly problematic. This ability to cope within a certain range of conditions is referred to as the coping range (Figure 2). Extremely dry or extremely wet years fall outside this coping range, but they have tended to occur quite infrequently. This system is vulnerable to conditions falling outside the coping range. The system has management options or adaptations to deal with modest droughts, for example, but severe droughts exceed the adaptive capacity of the system. While the boundaries of the coping range, also called coping thresholds or adaptation thresholds, may be portrayed as discrete values, in practice they are likely to be gradations, becoming more problematic for the system the greater the deviation from the norm. The shading in Figure 2 represents this gradation of vulnerability. It is also likely that this vulnerability gradient is nonlinear. Beyond a certain level of severity, vulnerability may increase exponentially (e.g., in case of systems susceptible to famine) or it may tend to level off (e.g., in systems where insurance kicks in).

Figure 2 also shows the way in which climate change relates to the vulnerability and adaptive capacity of a system. While the average drought severity has increased over the time period ($\overline{X}_p$ to $\overline{X}_f$), even at the end of the period the average condition ($\overline{X}_f$) is still within the coping range. The system has the capacity to deal with a changed average year. However, with a shift in average (yet with no change in the variability), there are also changes in the frequency and magnitude of extremes. In this case, it is the severe droughts that become frequent. The severe wet years become less frequent. Without some adaptation or adjustment, and some shift in the coping range, the system is more frequently exposed to severe droughts and hence is more vulnerable.

The relationship between a change in average conditions and variability and extremes is also represented in Figure 2b, where drought severity is portrayed as a normal (or Gaussian) distribution. Whereas severe droughts currently occur roughly one year in 20, with the shift in mean (climate change) and with no change in variability (the shape and variance of the distribution remain stable), severe droughts have approximately a 1 in 5 probability of occurring. The main point here is that climate change can alter the frequency and magnitude of extreme conditions, even without a change in variability.

Figure 3 shows that changes in the frequency and magnitude of extreme conditions can also change without a change in average but with a change in variability (Figure 3b) and with changes in both mean and variability (Figure 3c).

These ideas are not new. The relationships between climate change, the frequency of extremes, and coping ranges, "damage thresholds", "bands of tolerance", "bands of adaptation and adjustment" and vulnerability have been explored in somewhat similar forms by Hewitt and Burton (1971), Fukui (1979), Mearns et al. (1984), de Vries (1980, 1985), Hare (1985), Heathcote (1985), Wigley et al. (1985), Parry (1986), Warrick et al. (1986) and Handmer in Chapter 4 of this volume.

Thus, adaptation to climate change necessarily includes adaptation to climatic variability and extremes (Downing et al., 1996; Smit et al., 1999). Mileti (1999) and others use the term "climate hazards" to capture those climatic stimuli to which a system is vulnerable. Some systems such as low-lying islands or communities on permafrost are vulnerable to changes in average temperature. For most systems and communities, changes in mean conditions commonly remain within the coping range (the dark area on Figure 2). On the other hand, changes in the frequency and magnitude of extremes, associated with variability superimposed on a changing mean, may result in a system having to deal more frequently with conditions beyond its coping range (Rayner and Malone, 1998; Kelly and Adger, 2000; Chapters 2, 10 and 12 of this volume).

The coping range reflects a system's adaptive capacity. Many social and economic systems—including agriculture, forestry, settlements, transportation, and water resource management—have evolved to accommodate some deviations from "normal" conditions. This ability to deal with variations in climatic conditions from year to year up to some level may also be described in terms of a "vulnerability threshold". But rarely have these systems evolved to accommodate extremes.

While this range or threshold may be a useful concept to help understand vulnerability, it is not proposed that it is a discrete critical value that remains static over time. For example, a community may be somewhat vulnerable to a drought of a given severity and extent, but may be much more vulnerable to repeated droughts or droughts of greater severity or extent. It is unlikely that the "damages" or risks are linear with drought severity or frequency, nor are they likely to be discrete. It is also possible that another community in the same region may not be vulnerable to droughts at all, but to other climate

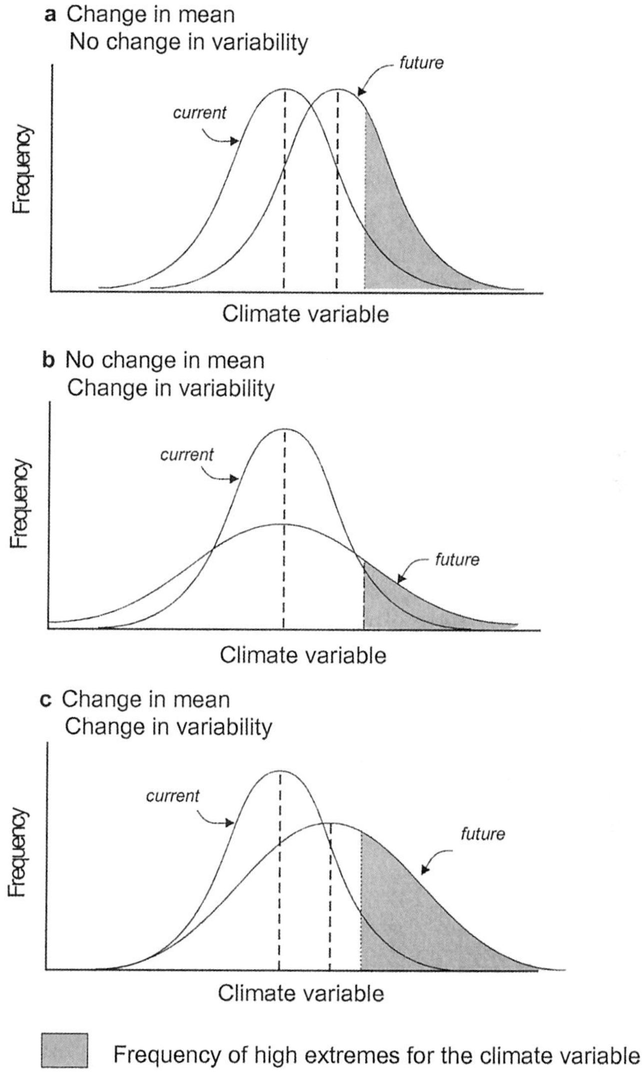

Figure 3. Changes in climate mean and variability (after Hare, 1985; Parry, 1986; Smit et al., 2000)

stimuli. The coping range is location-specific, group-specific and time-specific.

Of course, adaptive capacity varies from country to country, region to region and among social groups. It can also vary from household to household. Furthermore these scales of adaptive capacity are not independent. A household's capacity to deal with climate risks will depend, at least in part, on the enabling environment of the community within which the household functions. Similarly, a community's coping ability is rarely independent of the adaptive capacity of the region or state of which the community is part. Adger (Chapter 3 in this volume) demonstrates that adaptive capacity is place- and culture-specific and can only be identified through place- and culture-specific research.

The coping range of any system (household, community, nation) can also vary over time (deVries, 1985), with changes in economic, social, political and institutional conditions. Figure 4 illustrates conceptually several ways in which a coping range might change over time. In some places, population pressure or resources depletion may gradually reduce a system's flexibility and narrow its coping range (Figure 4a). Elsewhere, economic growth, improvements in technology and strengthening of social networks may lead to a gradual increase in adaptive capacity (Figure 4b). A more sudden decrease in political stability and the consequent problems with infrastructure and economic conditions may be reflected in a more sudden reduction or shift in coping ability (Figures 4c, 4d). It is also possible with targeted initiatives to move adaptive capacity in the direction in which the risks are moving (Figure 4e).

Any expansion of the adaptive capacity of a system or a community will reduce the vulnerability of that system to climate change risks, including those associated with deviations from average conditions. In addition, improvements in adaptive capacity invariably reduce vulnerability to present climate risks (see Chapters 4 and 7 of this volume). Development initiatives that increase capacity are essentially expanding the coping range.

The recognition that extremes are part of climate change, and that improvement of adaptive capacity is an effective means of reducing vulnerability to climate risks, has brought the climate change issue more predominantly onto the policy agenda of many countries (Downing, 1991; Burton, 1996; Huq et al., 1999; Chapter 7 of this volume). Adaptation is increasingly seen as an activity that can address potential impacts several decades hence, but also address climate-related hazards currently

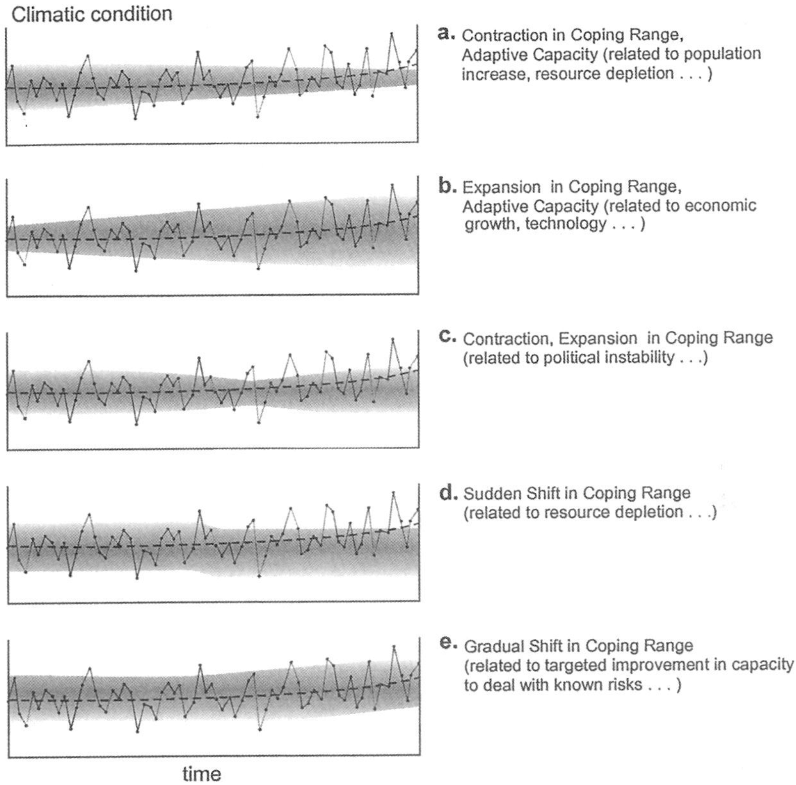

Figure 4. Changes in coping range/adaptive capacity (after deVries, 1985)

experienced with unfortunate frequency, particularly in the developing world (see Chapters 11 and 12).

3. Approaches to Adaptation Planning

The conventional approach to identifying adaptation options to climate change (Carter et al., 1994) has provided much knowledge on expected impacts of climate change, and has initiated consideration of the potential of adaptation measures to moderate adverse impacts (Smith, 1997; Tol et al., 1998; Klein et al., 1999; Leary, 1999; McCarthy et al., 2001). The approach broadly follows the sequence of steps apparent in Figure 1. Scenarios of future climate, often combined with scenarios of future socioeconomic conditions, are used to assess initial impacts or effects on the system, sector, region or community of interest. If autonomous adaptations are assumed, then residual or net impacts are estimated.

Where these net impacts are considered problematic, there is an interest in identifying particular activities, technologies or processes that would make the system better adapted to the expected climate scenarios. Once identified or postulated, these planned adaptation measures (Figure 1), sometimes called "micro-adaptation" (Summary chapter in this volume; Chapter 4 of this volume), might be subject to evaluation, prioritization and implementation. Given the resources and time needed to follow these steps, the adaptation component has only recently begun to receive serious attention (Burton, 1996; Klein, 1998). Much knowledge about the processes of adaptation has been found to be available from the fields of risk management, natural hazards research, and resource development and planning.

In exploring the possibilities for adaptation measures, it has become apparent that there are many different kinds of adaptation (Smithers and Smit, 1997; Smit et al., 2000). Table 1 shows how adaptations can be differentiated according to whether they occur in natural or human systems, are undertaken by governments or private interests, are autonomous or planned, are undertaken in advance or after impacts are experienced, are tactical or strategic, are local or broad in application, serve to retreat or protect or tolerate, take one of several forms and perform on any of numerous evaluative criteria. Adaptive responses can also be grouped according to the scale of implementation (Chapter 5 of this volume). These are all only generic differentiations. When adaptations are considered for a particular vulnerability or potential impact (for example, the vulnerability of Bangladesh communities to sea level rise and storms, or the vulnerability of pastoral agriculture in Argentina to changes in the frequency and severity of droughts) several realities quickly become apparent:

- Specific adaptation measures for a particular case are numerous. The types and combinations of options dealing with different kinds of structures, management practices, financial mechanisms and so on are almost infinite (Smit and Skinner, 2002). This can make the conventional process of identifying climate change impacts and adaptation options and evaluating their relative merit impractical.

Table 1. Bases for differentiating adaptations

Attributes	Examples of terms
System type	Natural—human Public—private
Purposefulness	Autonomous—planned Passive—active
Timing	Anticipatory—responsive Proactive—reactive
Temporal scope	Short term—long term Tactical—strategic
Spatial scope	Localized—widespread
Function/effects	Retreat—accommodate—protect Prevent—tolerate—spread—change—restore
Form	Structural—legal—institutional Regulatory—financial—technological
Performance	Cost-effectiveness—efficiency Implementability—equity

- Adaptation measures are very specific to a particular location and situation. What may work in one place or with one socioeconomic group may not work or may not be feasible elsewhere. Furthermore, there is a need to tailor adaptations to fit each case and circumstance. This means that it is essentially impossible to specify or develop appropriate and applicable adaptation measures without detailed knowledge of the system or community for which the adaptation is intended. This knowledge would need to include the nature of the vulnerability and how it is dealt with now to avoid proposing naively inapplicable, or even harmful or maladaptive, measures.
- Adaptation measures are unlikely to be implemented unless they are consistent with (or integrated into) broader management decision processes. Research and practice have shown that proposals to implement adaptive measures separately from existing policy and decision processes tend to go nowhere (Smit et al., 1996, Dolan et al., 2001). For decision makers in households, businesses, communities, organizations and governments, climate-related risks (vulnerabilities) will be considered and managed (adaptation) together with other risks, and usually within established risk management institutions, legal frameworks and decision processes (Chapters 4, 8 and 11).

These realities demonstrate the difficulties of attempting to reduce vulnerability to climate change by relying on the conventional approach of identifying specific adaptations, and then assessing and implementing adaptation measures. From the perspective of a planner or manager in a sector or community, there is a need to focus on climate change risks that are readily apparent, and to work on enhancing adaptive capacity overall rather than (or at least in addition to) focusing on particular climate change adaptation measures.

This approach is quite standard in the fields of hazard management and community and international development (Liverman, 1990; Blaikie et al., 1994; Downing, 1996). In these fields, successful projects deal with the pertinent risks, as identified by the community, and they aim to enhance the community's capacity to deal with risks, recognizing that the specifics will be handled at the local level. They also ensure that initiatives to improve adaptive capacity are incorporated into ongoing management processes.

These lessons are among the stimuli for the development of the vulnerability approach to adaptation to climate change that is increasingly being employed in climate change adaptation projects. The United Nations Development Programme (UNDP), the National Action Plans on Adaptation (NAPAs) and many other international development initiatives have adopted some form of the vulnerability approach.

4. Vulnerability and Adaptive Capacity

Analyses intended to help direct and facilitate adaptation initiatives or the improvement of adaptive capacity increasingly start with an assessment of the vulnerability of the system of interest to climate change (Chambers, 1989; Chapter 5 of this volume). Usually this involves identifying vulnerabilities to past and current climate, as well as how these risks might alter with climate change.

"Vulnerability assessment" is the inverse of "impact assessment" (Figure 1). Vulnerability is a property of a system relative to climatic conditions. Impacts are impacts of climate on a system, whereas vulnerability is vulnerability of the system to climate. Both are relative concepts, but the perspectives and starting points are different. The starting point for impact assessment is the stimulus (the specified climate, usually average conditions from a climate change scenario); the starting point for vulnerability assessment is the system (the community or region or sector). The climate attributes chosen are those to which the system is vulnerable (may be average or particular extremes).

There is now broad agreement that the vulnerability of a given system is related both to its exposure to climate change effects (sometimes called initial impacts) and to its capacity to deal with those effects (also called adaptability, coping ability and adaptive capacity). Generally, a system that is more exposed to a particular climate stimulus will be more vulnerable, and a system that has more adaptive capacity will tend to be less vulnerable because of the ability to moderate effects. This relationship is expressed formally:

$$V_{it}^{s} = f(E_{it}^{s}, A_{it}^{s}) \tag{1}$$

where

V_{it}^{s} = vulnerability of system i to climate stimulus s in time t

E_{it}^{s} = exposure of i to s in t

A_{it}^{s} = adaptive capacity of i to deal with s in t

While the functional relationship is not specified in Equation 1—its details will vary by context, location, sector, community and so on—it is understood that V is a positive function of E (the greater the exposure to climatic risks, *ceteris paribus*, the greater vulnerability) and a negative or inverse function of A (the greater the adaptive capacity, *ceteris paribus*, the less the vulnerability). Various models of vulnerability, based on this basic concept, have been proposed (Briguglio, 1995; Warrick et al., 1996; Gommes et al., 1998; Adger, 2000; Kelly and Adger, 2000; Moss et al., 2000; Bohle, 2001; Downing, 2001; Chapter 6 of this volume).

For vulnerability assessment, it is not possible or logical to consider climatic conditions and events separately from the characteristics of the system. Exposure, the first element in Equation 1, is a property of the system relative to climatic conditions. It is much like the concept of "natural hazards", where a climatic event is a hazard only if it is problematic or dangerous for the system (Blaikie et al., 1994; Handmer et al., 1999; Mileti, 1999). Exposure depends jointly on the characteristics of the system and on the attributes of the climate. To illustrate at an extremely coarse level, Switzerland and Bangladesh have different exposures to sea level rise, whereas Bangladesh and Florida have somewhat similar exposures to sea level rise. This is not just because of sea level rise, but also because of the location of the places relative to sea level. Of course, there are differences in exposure between Bangladesh and Florida because of the nature of human settlement in the two places. And within any place there are likely to be different exposures in different social groups. This concept of exposure is very similar to the "sensitivity" described in McMichael and Kovats (2000).

Adaptive capacity, the second element in Equation 1 is also a property of the system, referring to its ability to deal with the exposure or risk. It is equivalent to coping ability, and includes the capacity to prepare for, avoid or moderate, and to recover from exposure effects. Adaptive capacity may reflect resilience, stability, robustness, flexibility and other characteristics of a system (Smithers and Smit, 1997; Klein et al., 1998; Rayner and Malone 1998; Smit et al., 1999; Chapter 11 of this volume). Bangladesh and Florida may have somewhat similar exposures to sea level rise, but they have quite different adaptive capacities, and thus they have different vulnerabilities. Obviously, within any region, communities will have different exposures and different adaptive capacities, based on socioeconomic and other conditions, resulting in different vulnerabilities.

5. Enhancing Adaptive Capacity

For practical purposes, reducing vulnerability to climate change requires improving the capacity to adapt and/or reducing exposure. While there are opportunities for reducing exposure (e.g., development control or resettlement), in many countries these opportunities are limited. Enhancing adaptive capacity becomes a priority. This model recognizes that adaptation is less about identifying and implementing specific climate change adaptation measures and more about strengthening an ongoing process wherein resources are available to identify vulnerabilities and employ adaptive management strategies, appropriate for that context, to deal with climate risks along with other risks (Handmer et al., 1999; Chapter 7 of this volume).

Considerable attention has been given to the characteristics of systems (communities or regions) that influence their propensity or ability to adapt, or their ability to incorporate climate adaptation in policy development. Among these so-called "determinants of adaptive capacity" are economic wealth, technology, information and skills, infrastructure, institutions, social capital and equity (Smit and Pilifosova, 2001; Chapters 3 and 6 of this volume). While these can be considered separately, in practice they are closely interconnected.

Whether expressed as economic assets, capital resources, financial means, wealth or poverty, the economic condition of nations and groups is a major determinant of adaptive capacity (since economic resources facilitate preparation and recovery) (Burton and Van Aalst, 1999; Kates, 2000; Munasinghe and Swart, 2000). Lack of technology can impede adaptability by limiting the range of possible adaptation strategies, many of which are technology dependent (Goklany, 1995; Scheraga and Grambsch, 1998, Chapter 13 of this volume). Successful adaptation also requires a recognition of the necessity to adapt, knowledge about available options, the capacity to assess them and the ability to implement the most suitable ones (Fankhauser and Tol, 1997), all of which rely on the wide availability of information and skills.

Ability to adapt is often dependent on the effectiveness of social networks and infrastructure (Toman and Bierbaum, 1996; Chapter 3 of this volume). Similarly, adaptation is related to the stability and capacity of institutions to manage risks associated with climate and other conditions (Smith and Lenhart, 1996; Huq et al., 1999). It is also widely recognized that adaptive capacity is greater if social cultural and political institutions ensure that the allocation of power and access to resources are distributed equitably (Adger and Kelly, 1999; Handmer et al., 1999; Toth, 1999).

6. Vulnerability, Adaptive Capacity and Development

From the vast material dealing with adaptation, vulnerability and capacity enhancement reviewed in the IPCC TAR (Smit and Pilifosova, 2001), several messages are strongly apparent.

Reductions in the vulnerability of nations or communities to climate change require enhancements in their adaptive capacity. Promotion of specific ("micro") adaptation measures can help, but there are practical problems with this route and questions about its effectiveness. Working to improve the ("macro") adaptive capacity of a community or country has challenges also, but it may be as practicable, and it is likely to have a broader and more long-lasting effect on reducing vulnerability.

Actions to enhance adaptive capacity to climate change are often consistent with actions to promote development generally. Development programs have potential to improve adaptive capacity, yet they tend not to take risks of climate change into account (Burton and van Aalst, 1999). Furthermore, capacity building to reduce vulnerability to future climate change is likely to also reduce vulnerability to current climate risks. Yet it is unlikely that initiatives to adapt to climate will be undertaken separately from other resource management and development decisions. It is apparent that, to be effective, actions to promote adaptations or adaptive capacity need to be integrated into ongoing decision-making and policy processes.

Tailoring adaptation assistance to local needs requires the following actions:

- Address real local vulnerabilities, so that stakeholders buy into the issue and are interested in reducing vulnerabilities of which they are well aware.
- Involve real stakeholders early and substantively, so that any assistance is directed at known local vulnerabilities, and adaptation initiatives are realistic and designed to be consistent with existing institutions and decision processes.
- Connect with local decision-making processes, so that adaptation initiatives are developed relative to other conditions, are "mainstreamed" to the extent possible, and have the best possible chance of actually being implemented.

References

Adger, W. N. (2000), Social and ecological resilience: are they related? *Progress in Human Geography* **24**, 347-364.

Adger, W. N. and Kelly, P. M. (1999), Social vulnerability to climate change and the architecture of entitlements. *Mitigation and Adaptation Strategies for Global Change* **4**, 253-266.

Blaikie, P., Cannon, T., Davies, I. and Wisner, B. (1994), *At Risk: Natural Hazards, People's Vulnerability, and Disasters*. Routledge, New York.

Briguglio, L. (1995), Small island developing states and their economic vulnerabilities. *World Development* **23**, 1615-1632.

Burton, I. (1996), The growth of adaptation capacity practice and policy. In *Adapting to Climate Change: An International Perspective,* Smith, J., Bhatti, N., Munzhulin, G., Benioff, R., Budyko, M. I., Campos, M., Jallow, B. and Rijsbaerman, F., eds. Springer-Verlag, New York, 55-67.

Burton, I. and van Aalst, M. (1999), Come Hell or High Water: Integrating Climate Change Vulnerability and Adaptation into Bank Work. World Bank Environment Department Paper No. 72, Climate Change Series, Washington, DC.

Carter, T. R., Parry, M. L., Harasawa, H. and Nishioka, S. (1994), *IPCC Technical Guidelines for Assessing Climate Change Impacts and Adaptations*. University College, London, and Centre for Global Environmental Research, Tsukuba, Japan.

Chambers, R. (1989), Editorial introduction: vulnerability, coping and policy. *IDS Bulletin* **21**, 1-7.

De Vries, J. (1980), Measuring the impact of climate on history: the search for appropriate methodologies. *Journal of Interdisciplinary History* **10**, 599-630.

De Vries, J. (1985), Analysis of historical climate society interaction. In *Climate Impact Assessment: Studies in the Interaction of Climate and Society,* Kates, K.W., Ausubel, J. H. and Berberian M., eds. Wiley, Chichester, UK, 273-293.

Dolan, A. H., Smit, B., Skinner, M. W., Bradshaw, B. and Bryant, C. R. (2001), Adaptation to Climate Change in Agriculture: Evaluation of Options. Occasional Paper No. 26. Department of Geography, University of Guelph, Ontario, Canada.

Downing, T. E. (1991), Vulnerability to hunger in Africa. *Global Environmental Change* **1**, 365-380.

Downing, T. E., ed. (1996), *Climate Change and World Food Security.* Springer-Verlag, Berlin, Germany, 662.

Downing, T. E. (2001), *Climate Change Vulnerability: Linking Impacts and Adaptation*. Report to the Governing Council of the United Nations Environment Programme. Environmental Change Institute, Oxford, UK.

Downing, T. E., Olsthoorn, A. A. and Tol, R. S. J. (1996), *Climate Change and Extreme Events: Altered Risks, Socio-Economic Impacts and Policy Responses.* Vrije Universiteit, Amsterdam.

Fankhauser, S. and Tol, R. S. J. (1997), The social costs of climate change: the IPCC second assessment report and beyond. *Mitigation and Adaptation Strategies for Global Change* **1**, 385-403.

Fukui, H. (1979), Climate variability and agriculture in tropical moist regions. In *Proceedings of the World Climate Conference*. World Meteorology Report No. 537. Geneva, Switzerland.

Goklany, I. M. (1995), Strategies to enhance adaptability: technological change, sustainable growth and free trade. *Climate Change* **30**, 427-449.

Gommes, R. et al. (1998), *Potential Impacts of Sea-Level Rise on Populations and Agriculture.* Food and Agriculture Organization of the United Nations, Rome.

Handmer, J., Dovers, S. and Downing, T. E. (1999), Societal vulnerability to climate change and variability. *Mitigation and Adaptation Strategies for Global Change* **4**, 267-281.

Hare, F. K. (1985), Climatic variability and change. In *Climate Impact Assessment: Studies in the Interaction of Climate and Society,* Kates, K.W., Ausubel, J. H. and Berberian, M., eds. Wiley, Chichester, UK, 37-69.

Heathcote, R. L. (1985), Extreme event analysis. In *Climate Impact Assessment: Studies in the Interaction of Climate and Society,* Kates, K.W., Ausubel, J. H. and Berberian, M., eds. Wiley, Chichester, UK, 369-403.

Hewitt, K. and Burton, I. (1971), The Hazardousness of a Place: A Regional Ecology of Damaging Events. Department of Geography Research Paper No. 6, University of Toronto Press, Toronto, Canada.

Hulme, M., Barrow, E. M., Arnell, N. W., Harrison, P. A., Johns, T. C. and Downing, T. E. (1999), Relative impacts of human-induced climate change and natural climate variability. *Nature* **397**, 688-691.

Huq, S., Karim, Z., Asadazzaman, M. and Mahteb, F., eds. (1999), *Vulnerability and Adaptation to Climate Change in Bangladesh.* Kluwer, Dordrecht, The Netherlands.

Kates, R. W. 2000. Cautionary tales: adaptation and the global poor. *Climatic Change* **45**, 5-17.

Kelly, P.M. and Adger, W. N. (2000), Theory and practice in assessing vulnerability to climate change and facilitating adaptation. *Climatic Change* **47**, 325-352.

Klein, R. J. T. (1998), Towards a better understanding, assessment and funding of climate change adaptation. *Climatic Change* **44**, 15-19.

Klein, R. J. T., Nicholls, R. J. and Mimura, N. (1999), Coastal adaptation to climate change: can the IPCC technical guidelines be applied? *Mitigation and Adaptation Strategies for Global Change* **4**, 239-252.

Klein, R. J. T., Smith, M. J., Goosen, H. and Hulsbergen, C. H. (1998), Resilience and vulnerability: coastal dynamics or Dutch dikes? *The Geographic Journal* **164**, 259-268.

Leary, N. A. (1999), A framework for benefit-cost analysis of adaptation to climate change and climate variability. *Mitigation and Adaptation Strategies for Global Change* **4**, 307-318.

Liverman, D. M. (1990), Vulnerability to global environmental change. In *Environmental Risks and Hazards,* Cutter, S. ed. Prentice-Hall, Englewood Cliffs, NJ, USA, 326-342.

McCarthy, J. J., Canziani, O. F., Leary, N. A., Dokken D. J. and White, K. S. (2001), *Climate Change 2001: Impacts, Adaptation, and Vulnerability—Contribution of Working Group II to the Third Assessment Report of the Intergovernmental Panel on Climate Change.* Cambridge University Press, Cambridge, UK.

McMichael, A .J. and Kovats, R. S. (2000), Climate change and climate variability: adaptations to reduce adverse health impacts. *Environmental Monitoring and Assessments* **61**, 49-64.

Mearns, L. O., Katz, R. W. and Schneider S. H. (1984), Extreme high-temperature events: changes in their probabilities with changes in mean temperature. *Journal of Applied Meteorology* **23**, 1601-1613.

Mileti, D. S. (1999), *Disasters by Design: A Reassessment of Natural Hazards in the United States*. Joseph Henry Press, Washington, DC.

Moss, R. et al. (2000), Measuring Vulnerability: A Trial Indicator Set. Pacific Northwest National Laboratory, Richland, WA, USA.

Munasinghe, M. and Swart, R. eds. (2000), *Climate Change and its Linkages with Development, Equity and Sustainability, Proceedings of the IPCC Expert Meeting held in Columbo, Sri Lanka, April 1999.* LIFE, Colombo, Sri Lanka; RIVM, Bilthoven, The Netherlands; and the World Bank, Washington, DC.

Parry, M. L. (1986), Sustainable development of the biosphere: themes for research. In *Sustainable Development of the Biosphere*. Clark W. and Munn R., eds. Cambridge University Press, Cambridge, UK.

Rayner, S. and Malone, E. L. (1998), *Human Choice and Climate Change Volume 3: The Tools for Policy Analysis.* Battelle Press, Colombus, OH, USA.

Scheraga, J. and Grambsch, A. (1998), Risks, opportunities, and adaptation to climate change. *Climate Research* **10**, 85-95.

Smit, B. and Pilifosova, O. (2001), Adaptation to climate change in the context of sustainable development and equity. In *Climate Change 2001: Impacts, Adaptation, and Vulnerability—Contribution of Working Group II to the Third Assessment Report of the Intergovernmental Panel on Climate Change.* Cambridge University Press, Cambridge, UK, 876-912.

Smit, B. and Skinner, M.W. (2002), Adaptation options in agriculture to climate change: a typology. *Mitigation and Adaptation Strategies for Global Change* **7**, 85-114.

Smit, B., McNabb, D. and J. Smithers. 1996. Agricultural adaptation to climatic variation. *Climatic Change* **33**, 7-29.

Smit, B., Burton, I., Klein R. J. T. and Street, R. (1999), The science of adaptation: a framework for assessment. *Mitigation and Adaptation Strategies for Global Change* **4**, 199-213.

Smit, B., Burton, I., Klein R. J. T. and Wandel, J. (2000), An anatomy of adaptation to climate change and variability. *Climate Change* **45**, 223-251.

Smith, J. B. (1997), Setting priorities for adaptation to climate change. *Global Environmental Change* **7**, 251-264.

Smith, J. B. and Lenhart, S. S. (1996), Climate change adaptation policy options. *Climate Research* **6**, 193-201.

Smithers, J. and Smit, B. (1997), Human adaptation to climatic variability and change. *Global Environmental Change* **7**, 129-146.

Tol, R. S. J., Fankhauser, S. and Smith, J. B. (1998), The scope for adaptation to climate change: what can we learn from the impact literature? *Global Environmental Change* **8**, 109-123.

Tol, R. S. J., Jansen, H. M. A., Klein, R. J. T. and Verbruggen, H. (1996), Some economic considerations on the importance of proactive integrated coastal zone management. *Ocean and Coastal Management* **32**, 39-55.

Toman, M. and Bierbaum, R. (1996), An overview of adaptation to climate change. In *Adapting to Climate Change: An International Perspective,* Smith, J., Bhatti, N., Menzhulin, G., Benioff, R., Budyko, M. I., Campos, M., Jallow, B. and Rijsberman, F., eds. Springer-Verlag, New York, 5-15.

Toth, F. (1999), Development, equity and sustainability concerns in climate change decisions. In *Climate Change and its Linkages with Development, Equity and Sustainability: Proceedings of the IPCC Expert Meeting held in Colombo, Sri Lanka, 27-29 April, 1999,* Munasinghe, M. and Swart, R., eds. LIFE, Colombo, Sri Lanka; RIVM, Bilthoven, The Netherlands; and World Bank, Washington, DC, 263-288.

Warrick, R. A., Kenny, G. J., Sims, G. C., Erickson, N. J., Ahmad, Q. K. and Mirza, M. Q. (1996), Integrated model systems for national assessments of the effects of climate change: applications in New Zealand and Bangladesh. *Water, Air, and Soil Pollution* **92**, 215-227.

Wigley, T. M. L., Huckstop, N. J., Ogilvie, A. E. J., Farmer, G., Mortimer, R. and Ingram, M. J. (1985), Historical climate impact assessments. In *Climate Impact Assessment: Studies in the Interaction of Climate and Society*. Kates, K.W., Ausubel, J. H. and Berberian M., eds. Wiley, Chichester, UK, 529-565.

3. Social Aspects of Adaptive Capacity

W. Neil Adger
Tyndall Centre for Climate Change Research and
CSERGE School of Environmental Sciences
University of East Anglia, Norwich, United Kingdom

This paper outlines the nature of the relationship between climate change risks and social and institutional constraints on adaptation. Although the capacity of individuals to adapt to climate change is a function of their access to resources, the adaptive capacity of societies depends on the ability to act collectively in the face of the threats posed by climate variability and change. Thus adaptive capacity, as an element of overall vulnerability of a society, can be illuminated through examining the institutions for resource management and their effectiveness, efficiency and legitimacy. I argue that the social capital constitutes social aspects of adaptive capacity. Social capital is made up of the networks and relationships between individuals and social groups that facilitate economic well-being and security. Indeed I demonstrate that social capital is an important element for coping with climate variability and hazard in the present day. The paper concludes with challenges for research to bring social aspects such as legitimacy and equity to the heart of the analysis of adaptation.

1. Introduction

Societies have inherent capacities to adapt to climate change. These capacities are bound up in the ability of societies to act collectively. Decisions on adaptation are made by individuals, groups within society, organizations and governments on behalf of society. But all decisions privilege one set of interests over another and create winners and losers. Examining the social dynamics and outcomes of adaptation moves beyond simply accounting for the economic costs and benefits of adaptation. We need to consider the social acceptability of adaptation options, the institutional constraints on adaptation and the place of adaptation in the wider landscape of economic development and social evolution of societies into the future. So it is necessary first to distinguish adaptation by who is undertaking it, to what they are adapting and the embedded nature of social adaptation processes.

It is clear that individuals and societies have been adapting to climate change throughout human history and will continue to do so—climate is part of the wider environmental landscape of human habitation (de Menocal, 2001; Adger and Brooks, in press). Thus individual and societies are vulnerable to climate risks and other factors, and this vulnerability can act as a driver for adaptive resource management. Various scales and actors and involved in adaptation. Some adaptation by individuals is undertaken in response to climate threats, often triggered by individual extreme events (Ribot et al., 1996). Other adaptation is undertaken by governments on behalf of society, sometimes in anticipation of change but again, often in response to individual events. To understand the social aspects of adaptation related to climate variability and change, greater insights can be gained from looking around and looking back than from looking forward. In other words, we can observe directly both the patterns of adaptation in the past and today and the social and economic constraints on their efficiency and success. This is a necessary first step in theorizing the nature of adaptation processes. Explanations of past adaptation to climate change, and environmental change in general, have variously been based on theories of rational action, natural selection, and historical determinism, and all have been implicit within previous analyses of adaptation. Only with understanding of the assumptions of the underlying models of adaptation can the modeling of future adaptation options better reflect social evolution.

But there are limits to understanding the social dimensions of adaptation from past experience and perception in the context of climate change. Two are apparent. First, adaptations to present climate variability tend to be made within present day coping ranges. Positive strategies of adaptation often aim to expand the coping range of societies. Governments, for example, invest in dam construction to assure urban water supply, or implement inoculation campaigns. These investments always have socially differentiated impacts—they benefit some sections of society to the exclusion of others—farmers benefit from dam building, but others are displaced and lose out, for example. Some adaptation beneficial for climate change may therefore actually occur as acts of resistance to government policies or infrastructure schemes.

It is also unclear in looking at adaptation strategies today if the same unenlightened development strategies based on large-scale infrastructure and unsustainable development can be avoided in the future. So a second lesson from past adaptation is that although adaptive capacity may be high within the present coping range, future adaptation strategies may involve technological and institutional lock-ins that may, in effect, reduce the scope for increasing the range of adaptation options available. Thus avoiding maladaptation would seem to be as significant a challenge as implementing positive adaptation strategies.

I argue in this paper that levels of decision-making in society are not independent—they are embedded social processes that reflect the relationship between individuals, their networks, capabilities and social capital, and the state. Often a distinction is drawn between planned adaptation by governments on behalf of society and autonomous adaptation by individuals (summarized in Smit et al., 2001; see also Adger et al., in press). But this distinction obfuscates the role of the state in promoting development paths that cause institutional and technological lock-in and unsustainability. Khan, in Chapter 13, illustrates these points with examples of increased vulnerability as a result of government policy. Action is clearly required to enhance the adaptive capacity of the most vulnerable societies and groups that are often invisible to development and globalization processes. Resilience of those most vulnerable societies is dependent on the social capital of societies, the flexibility and innovation in the institutions of government and the private sector to grasp opportunities associated with climate change, and the underlying health and well-being of individuals and groups faced with the impacts of climate change.

This paper outlines the nature of vulnerability and the major social issues of conflict between adaptation and societal goals. It then outlines some emerging approaches within the social sciences that explain and describe the social aspects of adaptive capacity—namely the ability to act collectively and social capital. The paper also presents examples of adaptive capacity from existing studies of coping with climate extremes or managing vulnerable resources and draws out the research implications of the social aspects of adaptation.

2. The Relationship between Vulnerability and Adaptive Capacity

Social vulnerability is a state. It is the exposure of groups or individuals to stress as a result of social change and environmental hazard and change. This definition emphasizes the social dimensions of vulnerability following the tradition of analysis of vulnerability to hazards and food insecurity and as a dimension of entitlement to resources (see also Chapter 5). Handmer, in Chapter 4, discusses the interaction of vulnerability to climate change with vulnerability to hazards, arguing that few countries are currently well adapted and that today's vulnerable groups will only be more marginalized and more vulnerable under climate change stresses. The social turn in vulnerability analysis complements rather than replaces the traditional emphasis on physical vulnerability to the impacts of climate change. Vulnerability is therefore made up of a number of components, including exposure and sensitivity to risk and the capacity to adapt (following Brooks and Adger, Measures of observed vulnerability to climate-related discrete natural disasters, unpublished paper, Tyndall Centre for Climate Change Research, University of East Anglia):

Vulnerability = f(exposure, sensitivity, adaptive capacity).

For any given social and economic system, exposure is the probability of the occurrence of an extreme event whose influence extends over a particular area with particular characteristics. These characteristics include magnitude, frequency, duration, geographical extent, speed of onset, spatial dispersion and temporal spacing (see, e.g., Burton et al., 1993). Sensitivity is the extent to which a human or natural system can absorb the impacts without suffering long-term harm or some significant state change. This concept of sensitivity, closely related to resilience, can be observed in physical systems with impact-response models, but requires greater interpretation in ecological and social systems, where the definitions of harm and equilibrium change are more contested (Adger, 2000b). So adaptive capacity, in the context of this paper, is therefore a component of vulnerability, defined as the ability of a system to evolve in order to accommodate perturbations or to expand the range of variability with which it can cope (e.g., Jones, 2001).

There are generic features of adaptive capacity of societies to climate variability and change as well as to other types of stress. These have to do with the resources available to cope with exposure, the distribution of these resources across the landscape and between groups within a population, and the institutions that mediate both resources and coping with risk. Change in social vulnerability from its baseline level incorporates notions of economic development, adjustments to livelihoods based on adaptation to risk and changes in institutional and political structures. If institutions fail to plan for changing environmental conditions and risks, social vulnerability increases (Adger, 1999; 2000a).

A key lesson from much of the work on vulnerability to date is that it is socially differentiated. In other words, virtually all climate change differentially affects different groups in society depending on their ability to cope. The papers on coastal risks by Perez (Chapter 10) and agricultural adaptation and vulnerability by Conde and Eakin (Chapter 11) demonstrate the intricate and dynamic nature of these differential impacts over time. For flooding and land use change associated with climate change, the vulnerability of human populations is based on where they reside, their use of the natural resources and the resources with which they can cope. Many comparative studies point out that throughout human history the poor and marginalized have been most at risk from climate-related hazards (e.g., Hewitt, 1997).

Present and future vulnerability has strong social elements because it is a function of adaptive capacity, which is in turn dependent on social capital, institutions, resources and their distribution. Adaptive capacity is only potential until there are governance institutions that make it realizable. These institutions need legitimacy and harmony with wider social goals. I have argued elsewhere (Adger, 2001) that the appropriate scale of governance for adaptation is in fact the lowest subsidiary level. Sustainable resource management in effect requires empowered collective decision-making governance structures (Brown et al., 2002). Ostrom et al. (1999) argue that although the scale of many environmental problems is now global and that global action is required, simply replicating local-level collective action at the global scale is neither feasible nor appropriate given the need for unanimous agreement as a collective choice rule for global public goods and the diversity of cultural settings. In effect the diversity of impacts of climate change means that the most appropriate adaptation responses will often be multilevel responses.

This discussion of the appropriate scale of institutions to promote adaptation suggests that broader principles of sustainable development that promote "equality in opportunity" to adapt are required. The key problem is

Table 1. Adaptation pathways and strategies and social dimensions of their implementation

Adaptation pathways and strategies	Social dimensions necessary for implementation
Urban planning and zoning to avoid flood, heat stress and other hazards.	Community governance and participatory structures. Effective means of dealing with social exclusion and urban underclasses.
Planning for demographic transition and regional planning.	Social mobility, coherent regional identity, social tolerance and mixing.
Large-scale infrastructure development (e.g., dams, irrigation and water management facilities).	Social acceptance of apparently 'modernist' development ideas that have had detrimental social impacts to excluded groups in the past.
New technologies in agriculture and natural resource use.	As with above, social acceptance of technologies. Recognition of existing non-scientific knowledge and technologies.
Policies and plans for natural areas.	New institutional structures for protected area management overturning past models of national parks and exclusion of people-centered-conservation.
Mitigation policies in energy, transport and other sectors.	Social acceptance of winners and losers in new technologies, fiscal and regulatory policies and other impacts.

the identification of adaptation options that are most harmonious with existing social norms, institutions and structures. Table 1 outlines a number of adaptation pathways and options and suggests some well documented social dimensions of adaptation planning within these areas. So, for example, urban planning and land use zoning in general take place within local governance structures. But the enforcement and effectiveness of planning and zoning depend on the inclusionary and consensual nature of the processes. Often key vulnerable groups are excluded. Poorer households are forced to live in riskier areas in urban settlements, putting them at risk from flooding for example. And they are frequently ignored when designing infrastructure to alleviate such vulnerabilities. In the words of Scott (1998), the space occupied by socially marginalized groups becomes itself invisible. This has been shown in many urbanized areas, in Japan (Uitto, 1998), the United

States (Cutter al., 2000) and Guyana (Pelling, 1999), for example. Exclusion for decision-making is often based on social status or other factors such as gender and age (e.g., Fordham, 1999). When collaborative planning is ignored the sustainability of plans and their implementation is in question (e.g., Healey, 1997).

These observations on the social dimensions of adaptation strategies in urban planning also hold true for proposed new technologies in agriculture and other areas—social and environmental movements have sprung up because of social mistrust and disquiet with genetically modified crops. These technologies cannot be simply assumed as potential adaptation strategies for drought resistance or yield security. There is also, for example, increasing disquiet concerning the social impacts of protecting areas around the world for conservation by excluding all human use. There is little point therefore of planning for new protected areas to promote conservation in the face of new climatic conditions without confronting debates on people-centered conservation and new ways of integrating human use of landscapes (see, for example, Ghimire and Pimbert, 1997; Noss, 2001; Brown, 2002). And in the same vein, many climate change mitigation policies are opposed because of the assumed retrogressive nature of taxes or regulations on poorer sections of society. Of course, these perceptions of taxes hitting the poor hardest are often myths propagated by lobby groups who wish fossil fuel use to remain subsidized or unregulated (see Tindale and Hewett, 1999). So for all the adaptation strategies highlighted in Table 1, their social acceptance and harmony with wider sustainable development goals and aspirations is critical. Ignoring these dimensions of adaptation, as I argue below, reduces their effectiveness, efficiency and legitimacy. Societies adapt to climate change through collective action, mediating and trading off the elements of effectiveness and legitimacy through negotiated outcomes.

3. Collective Action and Social Capital

In the previous section I argued that collective action is at the heart of all adaptation decisions. Collective actions are those taken by a group (either directly or on its behalf through an organization) in pursuit of members' perceived interests whereby mutual benefits can be gained by individuals cooperating rather than working against each other to achieve an outcome. In agriculture, forestry and other resource-dependent livelihoods, resources frequently exist under multiple property rights regimes, there are many different users and there is limited information about the impacts of environmental change on sustainability. Diverse social sciences from anthropology to psychology explore how societies make choices about allocating scarce resources to provide an understanding of choices despite limited information and uncertain futures. These theories are distinct and often in conflict about methods, scope and the framing of the questions—i.e., Whose decisions? And what decisions? Thus the means of decision-making and the outcomes of decision-making, from efficiency, equity and legitimacy perspectives, are all contested concepts (Adger et al., 2002). Collective action requires networks and flows of information between individuals and groups to oil the wheels of decision-making. These sets of networks constitute, in effect, social capital.

Social capital is therefore made up of "the norms and networks that enable people to act collectively" (Woolcock and Narayan, 2000). These networks are scale dependent and exist both outside the state and between the state and other elements of society. Social capital is associated with a progressive and perhaps flexible and adaptive society. It is a necessary "glue" for social resilience, but also for economic development (Adger, 2000b). The prevalence of different types of social capital, however, is important at different times to different social groups. It is made up of the sharing of knowledge, sharing of financial risk, sharing market information, or claims for reciprocity in times of crisis. Ties within a defined socioeconomic group have come to be known as *bonding social capital* and may be based on family kinship and locality. By contrast *networking (or bridging) social capital* is made up of the economic and other ties external to social groups.

In addition to the institutions of governance and their rules, "community" is the most important ingredient for collective management (discussed extensively in Brown et al., 2002). Informed and empowered communities can create the environment for voluntary action to provide social order, to resolve conflicts and to regulate common resources. If hybrid co-management between governments and communities is to succeed, then these entities must mobilize and foster cooperative relations among different social groups. The diversity of interests within "communities" involved in managing resource subject to climate stress and the wider democratic framework within which such management occurs mean that, in effect, "community" management is not always effective (Hayward, 1995).

Although social capital has been widely promoted as a way to understand the functioning of society, there are divergent views in two main areas—whether social capital exists only outside the state and whether it is a cause or simply a symptom of a progressive and perhaps flexible and adaptive society. Each of these controversies is important for adaptive capacity for climate change. The first issue, on the importance of the state, relates to the importance of planned adaptation strategies. In other words, if governments can provide physical infrastructure to minimize the potential impacts of floods or droughts, will this be sufficient for adaptation if its use does not resonate with social norms? The second issue of the macro-level functioning of society is again important for adaptive capacity and an area in which there are initial steps toward quantifying the role of social capital (Knack and Keefer, 1997).

Social capital is made up of the sharing of knowledge, sharing of financial risk, sharing of market information, or claims for reciprocity in times of crisis. Ties within a defined socioeconomic group have come to be known as bonding social capital and may be based on family kinship and locality. Networking (or bridging) social capital is made up of the economic and other ties external to the group. Networking social capital tends to be based not on friendship and kinship but on the weaker bonds of trust and reciprocity. Hence networking social capital has less formal collective action rules of enforcement and sanction and becomes increasingly reliant on legal and formal institutions.

The importance of the investigation of the role of social capital in economic activities is that it moves household economic models away from the notion that social relations simply muddy the waters for efficient economic transactions and rational resource allocation. This notion, that social relations always constitute a constraint, has long been questioned by political scientists (see, e.g., Scott, 1976) who have shown that the patron-client relations in agrarian societies represent a moral economy often in operation at times of hardship. Such peculiar and specific moral economies are often a necessary component of coping with weather extremes and other hazards and their impacts (Ribot et al., 1996; Pelling, 1999).

But this does not mean that more networks, greater reciprocal relations and commitments and generally "more" social capital is always beneficial for all individuals or all situations. There is, in effect, a dark side to the force. Economists argue that excessive bonding social capital, while particularly necessary for poorer sections of society can constrain economic development. It is the different combinations of bonding and networking social capital that allow communities to confront poverty and vulnerability, resolve disputes and take advantage of new opportunities (Woolcock and Narayan, 2000). Similarly not all social networks can be harmonious with good governance and the operation of society—as Woolcock (1998) points out, criminal organizations have strong social capital but their objectives subvert the social capital of others in society and rather may constitute "social disorganization" (Arrow, 2000).

Bridging or networking social capital, based on informal institutions but relying on legal sanctions, is more akin to what we normally perceive as common property management regimes. Community groups, particularly but not exclusively in developing countries, provide basic health care and manage weather-related risks. They undertake resource management of fisheries, forests and irrigation systems and use the institutions to resolve disputes and allocated benefits (reviewed extensively in Baland and Platteau, 1996). It is clear how networking social capital is important to adaptive capacity in the context of climate change—through managing risk. These are discussed in the example from Vietnam in Section 4.

But bonding social capital has also been recognized as an important component of coping with the adverse impacts of climate change. Epidemiologists concerned with environmental triggers to ill health also necessarily include the social context and characteristics of both populations and individuals in their analyses. Table 2 reports results of research into over 300 cases among the estimated 700 excess deaths in Chicago as a result of a heat wave in July 1995 (Semenza et al., 1996). It shows, by calculating the odds ratios of being a case compared to a control sample, the impact of various socioeconomic characteristics of the cases. The results show that, leaving aside pre-existing medical risks, the factor most significantly increasing the risk of mortality is location (living in the top floor of non-apartments or in hotels) along with absence of working air-conditioning. But also increasing the risk of mortality is whether the individuals lived alone and stayed at home through the heat wave. Social interaction, or bonding social capital, decreases the risk of mortality from cardiovascular related illness in these cases. Attending a church or social group, or visiting friends may not in themselves enhance the ability of at risk groups to cope with heat waves, but clearly they are important for collective coping with such stresses. In his book on the decline of democratic civil society in the United

Table 2. Dying alone: socioeconomic status and social contacts of cases of heat related deaths during the 1995 heat wave in Chicago

Variable	Odds ratio of risk of death (95% CI)
Living conditions	
Lived on top floor	4.7 (1.7–12.8)
Lived alone	2.3 (1.4–3.5)
Visited cooling shelters	0.5 (0.3–1.2)
Working air-conditioner	0.2 (0.2–0.4)
Residence type	
Single family home	1.0
Apartment building	2.5 (1.5–4.2)
Other kind of building	8.1 (1.4–45.8)
Lived in building with flat roof	2.0 (1.2–3.2)
Social contact	
Did not leave home	6.7 (3.0–15.0)
Participated in group activities (church, support groups)	0.7 (0.5–0.9)
Had pet in home	0.6 (0.4–0.9)
Access to transportation	0.4 (0.2–0.6)
Had friends in Chicago	0.3 (0.1–0.6)

Notes: Odds ratios are calculated as the risk of death among subjects with the characteristic in question compared to those without it for 339 matched pairs of case subjects and controls. The July 1995 heat wave resulted in over 700 excess deaths classified as heat-related.

Source: Adapted from Semenza et al. (1996).

States, Putnam (2000) invokes the metaphor of citizens "bowling alone" where in past decades they participated in bowling clubs. Similar engagement in civil society reduced the risk of Chicago citizens from dying alone.

Can social capital be encouraged for legitimate and equitable adaptation by governments? An institutional view of social capital argues that "the very capacity of social groups to act in their collective interest depends on the quality of the formal institutions under which they reside" (Woolcock and Narayan, 2000, p. 234). Clearly the role of government in encouraging the development of social capital and local institutions are intertwined. Governments may not always be in a position to provide adequate resources and infrastructure for adaptation. In such circumstances and for the sake of effectiveness, they need to undertake adaptation in partnership with civil society.

There are some types of adaptation to climate change (e.g., provision of large-scale infrastructure) that are only easily realizable by the state. A well-functioning government can provide the necessary underpinning and social security for marginalized groups. The idealized situation is a synergy between state and civil society (Evans, 1996) that promotes social and policy learning. Social learning is a composite of individual adaptation, and is that part of adaptation that comes about through collective activities such as discourse, imitation and conflict resolution. It should be stressed that collective action is not necessarily for everyone's benefit. Social hierarchies and resource and entitlement inequalities are rarely overturned in the course of adaptation, and external changes such as climatic extremes and other natural hazards tend to reinforce those inequalities (Adger et al., 2001). As individuals and group interact synergistically with the state, so too do the institutions of the state evolve in a process often characterized as policy learning. Adaptation in the political sphere involves periodic shocks to ideologies and paradigms of policy intervention such that these external shocks are conduits of social learning and adaptation.

New institutions, even those instigated by the state, can build trust and legitimacy. In the United Kingdom inclusionary and participatory planning processes for coastal defense are novel but increasingly utilized. They promote long-term decisions on adaptation when there are urgent and divisive conflicts over short-term goals and interests. O'Riordan (2001) and O'Riordan and Ward (1997) report on the building of trust between local stakeholders and government agencies in planning for setback of the East Anglian coast. In this case both nature conservation organizations and local economic interests came to a compromise in designing coastal protection schemes. Social learning in this case came about through the widespread recognition and validation of perceptions of risk of local stakeholders.

But when a state is driven by ideology, colonialism or other circumstances to be in conflict with civil society, conflict ensues and the most marginal sections of society are made vulnerable. These circumstances are all too common. Civil strife and population displacement often trigger famine even in the absence of a fall in food production or an environmental catastrophe. This situation explains some of the major famines of the 20th century, including those in India and China in the 1940s and 1950s (Nolan,

1993). In the absence of an effective state, networking social capital is forced to substitute for some or many of the roles provided by governments. But the outcomes are often far from desirable.

A summary of the issues raised is presented in Table 3. This outlines the pertinent features (following Woolcock and Narayan, 2000) of the horizontal networking approaches to social capital, the institutional approaches, and the synergistic approaches and their implications for adaptive capacity to climate change. First the network view assumes that social capital is a phenomenon outside of the reaches of the state—social capital often substitutes for state involvement in the provision of public goods and is synonymous with the idea of civil society. Institutional approaches to social capital emphasize structure rather than agency. These insights are also used in comparative analysis of economic performance. Political freedom, bureaucratic performance and others are readily measurable at the cross-country level, and more easily quantifiable than civil society phenomena (Brooks and Adger, Measures of observed vulnerability to climate-related discrete natural disasters, unpublished paper, Tyndall Centre for Climate Change Research, University of East Anglia). The synergistic approaches to social capital recognize the limitations of both but are focused less on measuring the presence, absence or density of social capital than on the processes by which state and civil society interact through their embeddedness and complementarity (Evans, 1996).

These approaches are not mutually exclusive and all have, I would argue, useful perspectives on adaptive capacity in the area of climate change. Networking social capital is clearly important at the local level for social differentiation in vulnerability. Bonding social capital such as networks of reciprocity can clearly be an important asset for coping with the impacts of weather extremes and catastrophic events. In the discussions of network views of social capital, such bonding capital, although important for coping, does not necessarily facilitate proactive adaptation and enhancement of well-being.

Networked social ties to external agents are also clearly important both for coping and for evolutionary adaptation. For small island microstates, international migration assists both in both coping when extreme events occur and in furthering the stability and resilience of island populations. Many small island societies have proved resilient in the past to social and environmental upheaval (Barnett, 2001). The key parameters of this

Table 3. Views of social capital and implications for adaptive capacity in the context of climate change

Emphasis in social capital literature	Features applied to well-being and welfare	Implications for adaptive capacity
Networks	Stresses horizontal linkages and the role of non-state actors. The density of social capital leads to measurable outcomes in terms of material well-being.	Social capital of individuals and groups important for geographical and social manifestations of vulnerability and coping with risks.
Institutional approaches	Civil society operates to the degree that institutions of the state facilitate it.	The role of the state remains important for planned adaptation and sustainable development. Governance vital in managing global environmental risks and mitigative capacity.
Synergy approaches	Argues that state-society links and density are key. Addresses the complementarity and potential substitution of state and non-state and the normative issues of promoting environment for social learning.	State-society linkages important both for wider sustainable development and the co-management of resources. States can facilitate sustainable and resilient resource management and hence adaptive capacity.

resilience have included opportunities for migration and subsequent remittances but also traditional knowledge, institutions and technologies, subsistence economies adapted to climate hazards, and linkages between formal state and customary decision-making processes. This resilience may well be being undermined, however, as small island states increasingly become integrated into the world economy, in agriculture, tourism and mineral resources (Barnett and Adger, 2001) and where international migration is becoming less legitimate and more constrained (O'Neill et al., 2001). Migration strategies have been used throughout human history in such regions to promote resilience, though have often been seen through the lens of dependency and negative social and cultural impacts (Barnett and Adger, 2001). Thus social capital perspectives can have direct policy recommendations. Civil society institutions, although not specifically designed for the purpose, can play an important role in coping with the impacts of climate variability and change and can be encouraged through appropriate interventions.

But there are some public goods that can only easily be provided by the state, including major infrastructural changes necessary when the impacts of climate change are significant or are risky to large populations. The synergistic approaches to social capital suggest that implementing planned adaptation to climate change is best brought about by co-managing resources so that stakeholders from civil society buy into a shared vision of risk and adaptation in the long run and sustainable resource management in the immediate term. This issue is addressed in the case study of coastal management below.

4. An Example of the Evolution of Social Capital and Adaptive Capacity

To reiterate the central message, the social dynamics of adaptive capacity are defined through the ability to manage resources collectively. Resource-dependent communities have historically managed their resources in some collective fashion because of the nature of the fugitive, fluctuating and seasonal resources on which their livelihoods often depend. Synergistic social capital, where local management and government intervention work together to reduce risks, I would argue, promotes the adaptive capacity of societies to cope with climate change. But social capital is also fragile and can be lost or undermined by social and political change.

Like social and ecological resilience, social capital is often observable only when there is some perturbation to the social or ecological system in focus (see Carpenter et al., 2001). In the case of the impacts of hazards social networks play a primary role in coping and recovery. Where government intervention to plan for, forewarn and assist in recovery is largely absent, social capital, both bonding and networking, in effect takes over as a substitute for such external intervention. The rolling back of the state in times of crisis or "adjustment" presents communities with the necessity to provide alternative social security and other functions. Such rolling back of the state is evident in the post-socialist countries where many functions previously provided by the state have collapsed. This poses major challenges for managing resources and for the evolution of new institutions to provide social resilience (Adger et al., 2001). Suddenly in the mid-1990s the local-level hazard planning and coastal defense in Vietnam was confronted with decentralization and the break-up of agricultural cooperatives. The resulting institutional response proved to be an example of social capital substituting for the state (detailed in Adger, 2000a).

Sea dikes constructed for coastal defense in coastal northern Vietnam are the principal physical infrastructure investment to ameliorate the threat of climatic hazards associated with typhoons and coastal storms, and till the mid-1990s were the major responsibility of the coastal communes and districts. Agricultural cooperatives during the collectivization period had the responsibility for managing these defenses. Each adult allocated 10 days of labor to the task of repairing and maintaining the sea dike system. Since the decollectivization of agriculture, this role of the agricultural cooperatives has largely been made redundant and the sea defenses in many areas were for a number of years not being maintained, exacerbating vulnerability to present day climatic extremes.

Field research in 1996 and 1997 documented the new institutions and networks for collective maintenance of the defenses. Decentralized communes engaged in obfuscation and non-decision-making to divert remaining resources away from coastal defense to their higher priorities of aquaculture development. Thus the decentralization process, far from increasing local accountability, simply further exacerbated vulnerability to coastal storms.

But emerging social capital offset these trends toward greater vulnerability by creating networks for coping with present day weather extremes. The private property resolutions of Vietnam's 1992 Constitution, and the subsequent reform of the land allocation system under the 1993 Land Law, have been paralleled by a relaxation in the legality of private credit systems in rural areas. In the collectivized period, formal credit was permissible only through the commune cooperative. Informal, but illegal, credit systems have always been an integral part of the coastal fishing economy in this part of the world. This phenomenon is common to many artisanal fishing communities. The role of credit in recovery from stress and disruption of livelihoods is particularly important where external assistance is not available for immediate injection of resources. Street associations are informal associations of neighbors within hamlets who have traditionally maintained religious buildings, funeral and marriage ceremonies. Associations, along with reciprocal feasting and gift exchange, have become revitalized in northern Vietnam—it has long been recognized that these processes play a role in sustaining kinship and lineage ties that are necessary for security in times of crisis. Informal collective decision-making for coastal defense and new networking social capital in this instance substitute for the loss of state planning. Undoubtedly the adaptive capacity of many areas of the world when faced with climate changes will be to fall back on the local-level social networks. Hence the preconditions for the emergence and promotion of social capital remain an important element, and an ever-increasingly important element, in overall adaptive capacity.

5. Conclusions

A number of lessons for adaptation policy and analysis come from this rich agenda on collective action, social capital and adaptive capacity. First, the nature of adaptive capacity is such that it has culture- and place-specific characteristics that limit large-scale modeling and scaling up of models of adaptive capacity. Further, policy interventions for planned adaptation at national and other levels of efficient policy making may not be sensitive to these culture- and place-specific characteristics and hence adaptive capacity will be differentially affected by such policies. But this does not mean that the lessons from research on social capital are not generalizable. Rather it means that adaptive capacity can be mapped in terms of the prevailing governance structures rather than by focusing on social variables as is standard in social vulnerability analysis.

A second lesson is that in order to generalize we need to learn from theoretical insights into institutional form and in particular from research on the institutional prerequisites for the evolution and persistence of collective action and its relative importance compared to state intervention. From such literature it is clear that the size of the group undertaking the collective action, the boundaries of the resource at risk, homogeneity of the decision-making group, the distribution of benefits of management and other factors are all important in determining the ultimate success of collective management. Research is required on how collective action is central to adaptive capacity at various scales. Such research needs to be place and culture specific, and is amenable at all scales to diverse research approaches, from agent-based modeling to game theoretic approaches to qualitative assessments of the efficacy of rules and institutions (e.g., Becker and Ostrom, 1995; Agrawal, 2001). Insights in these areas, as provided in some of the papers in this book, represent a necessary building block of a theory of adaptation to climate change.

A third lesson is therefore that institutional theories of social capital provide a means to generalize on the macro-determinants of adaptive capacity. A large body of research argues that the state is central to the facilitation of social capital. Appropriate proxies for the performance of the state can be used to compare the prior conditions for adaptive capacity to climate change across states. This is the theoretical underpinning, though not always recognized, of a number projects seeking to identify the determinants of adaptive capacity at the national level (Moss et al., 2001; Brooks and Adger, 2002). The measurement and observation of social capital remain problematic. Bonding and networking social capital are not easily quantifiable phenomena. In many studies their presence or absence is approached through the number and extent of contacts, memberships and other proxies. At the macro-level there are more easily quantifiable proxies but these become more loosely correlated with the social capital phenomenon in question. Throughout this paper I have emphasized that activities to enhance social resilience and sustainable development in the present may not be obviously climate related. Climate change adaptation is a challenge to implementing sustainable and equitable development.

Acknowledgements

I thank Bob Kates and John Handmer for insightful comments on an earlier draft. This version is my own responsibility. I also thank Joel Smith and Richard Klein for encouragement and for inviting me to the workshop in Potsdam on which this book is based. The other participants also made it a stimulating occasion. Research funding by the Tyndall Centre for Climate Change Research is gratefully acknowledged.

References

Adger, W. N. (1999), Social vulnerability to climate change and extremes in coastal Vietnam. *World Development* **27**, 249-269.

Adger, W. N. (2000a), Institutional adaptation to environmental risk under the transition in Vietnam. *Annals of the Association of American Geographers* **90**, 738-758.

Adger, W. N. (2000b), Social and ecological resilience: are they related? *Progress in Human Geography* **24**, 347-364.

Adger, W. N. (2001), Scales of governance and environmental justice for adaptation and mitigation of climate change. *Journal of International Development* **13**, 921-931.

Adger, W. N. and Brooks, N. (in press), Does global environmental change cause vulnerability to disaster? In *Natural Disasters and Development in a Globalising World,* Pelling, M., ed. Routledge, London.

Adger, W. N., Kelly, P. M. and Ninh, N. H. (2001), *Living with Environmental Change: Social Vulnerability, Adaptation and Resilience in Vietnam*. Routledge, London.

Adger, W. N., Huq, S., Brown, K., Conway, D. and Hulme, M. (in press), Adaptation to climate change in the developing world. *Progress in Development Studies.*

Adger, W. N., Brown, K., Fairbrass, J., Jordan, A., Paavola, J., Rosendo, S. and Seyfang, G. (2002), Governance for sustainability: towards a "thick" understanding of environmental decision-making. Working Paper EDM 02-04. Centre for Social and Economic Research on the Global Environment, University of East Anglia, Norwich, UK.

Agrawal, A. (2001), Common property institutions and sustainable governance of resources. *World Development* **29**, 1649-1672.

Arrow, K. (2000), Observations on social capital. In *Social Capital: A Multi-Faceted Perspective*, Dasgupta, P. and Serageldin, I., eds. World Bank, Washington, DC, 3-5.

Baland, J. M. and Platteau, J. P. (1996*), Halting Degradation of Natural Resources: Is There a Role for Rural Communities?* Clarendon, Oxford, UK.

Barnett, J. (2001), Adapting to climate change in Pacific island countries: the problem of uncertainty. *World Development* **29**, 977-993.

Barnett, J. and Adger, W. N. (2001), Climate dangers and atoll countries. Working Paper 9. Tyndall Centre for Climate Change Research, University of East Anglia, Norwich, UK.

Becker, C. D. and Ostrom, E. (1995), Human ecology and resource sustainability: the importance of institutional diversity. *Annual Review of Ecology and Systematics* **26**, 113-133.

Brown, K. (2002), Innovations for conservation and development. *Geographical Journal* **168**, 6-17.

Brown, K., Tompkins, E. L. and Adger, W. N. (2002), *Making Waves: Integrating Coastal Conservation and Development*. Earthscan, London.

Burton, I., Kates, R. W. and White, G. F. (1993), *The Environment as Hazard*, 2nd edition. Guilford Press, New York.

Carpenter, S., Walker, B., Anderies, J. M. and Abel, N. (2001), From metaphor to measurement: resilience of what to what? *Ecosystems* **4**, 765-781.

Cutter, S. L., Mitchell, J. T. and Scott, M. S. (2000), Revealing the vulnerability of people and places: a case study of Georgetown County, South Carolina. *Annals of the Association of American Geographers* **90**, 713-737.

De Menocal, P. B. (2001), Cultural responses to climate change during the late Holocene. *Science* **292**, 667-673.

Evans, P. (1996), Government action, social capital and development: reviewing the evidence on synergy. *World Development* **24,** 1119-1132.

Fordham, M. (1999), The intersection of gender and social class in disaster: balancing resilience and vulnerability. *International Journal of Mass Emergencies and Disasters* **17**, 15-37.

Ghimire, K. and Pimbert, M. P. (1997), *Social Change and Conservation*. Earthscan, London.

Hayward, B. M. (1995), The greening of participatory democracy: reconsideration of theory. *Environmental Politics* **4**, 215-236.

Healey, P. (1997), *Collaborative Planning: Shaping Plans in Fragmented Societies*. Macmillan, London.

Hewitt, K. (1997), *Regions of Risk: A Geographical Introduction to Disasters*. Longman, Harlow, Essex, UK.

Jones, R. N. (2001), An environmental risk assessment/management framework for climate change impact assessments. *Natural Hazards* **23**, 197-230.

Knack, S. and Keefer, P. (1997), Does social capital have an economic payoff? A cross-country investigation. *Quarterly Journal of Economics* **112**, 1251-1288.

Moss, R., Brenkert, A. and Malone, E. L. (2000),Vulnerability to Climate Change: A Quantitative Approach. Report to the U.S. Department of Energy. Pacific Northwest National Laboratory, Richland, WA, USA.

Nolan, P. (1993), The causation and prevention of famines: a critique of A. K. Sen. *Journal of Peasant Studies* **21**, 1-28.

Noss, R. F. (2001), Beyond Kyoto: forest management in a time of rapid climate change. *Conservation Biology* **15**, 578-590.

O'Neill, B. C., MacKellar, F. L. and Lutz, W. (2001), *Population and Climate Change*. Cambridge University Press, Cambridge, UK.

O'Riordan, T. (2001), On participatory valuation in shoreline management. In *Economics of Coastal and Water Resources: Valuing Environmental Functions,* Turner, R. K., Bateman, I. and Adger, W. N., eds. Kluwer, Dordrecht, The Netherlands, 323-340.

O'Riordan, T. and Ward, R. (1997), Building trust in shoreline management: creating participatory consultation in shoreline management plans. *Land Use Policy* **14**, 257-276.

Ostrom, E., Burger, J., Field, C. B., Norgaard, R. B. and Policansky, D. (1999), Revisiting the commons: local lessons, global challenges. *Science* **284**, 278-282.

Pelling, M. (1999), The political ecology of flood hazard in urban Guyana. *Geoforum* **30**, 240-261.

Putnam, R. D. (2000), *Bowling Alone: The Collapse and Revival of American Community.* Simon and Schuster, New York.

Ribot, J. C., Magalhães, A. R. and Panagides, S. S. (1996), *Climate Variability, Climate Change and Social Vulnerability in the Semi-arid Tropics.* Cambridge University Press, Cambridge, UK.

Scott, J. C. (1976), *The Moral Economy of the Peasant: Rebellion and Subsistence in Southeast Asia.* Yale University Press, New Haven, CT, USA.

Scott, J. C. (1998), *Seeing Like a State: How Certain Schemes to Improve the Human Condition have Failed.* Yale University Press, New Haven, CT, USA.

Semenza, J. C., Rubin, C. H., Falter, K. H., Selanikio, J. D., Flanders, W. D., Howe, H. L. and Wilhelm, J. L. (1996), Heat related deaths during the July 1995 heat wave in Chicago. *New England Journal of Medicine* **335**, 84-90.

Smit, B. et al. (2001), Adaptation to climate change in the context of sustainable development and equity. In *Climate Change 2001: Impacts, Adaptation and Vulnerability*, McCarthy, J. J., Canziani, O., Leary, N. A., Dokken, D. J. and White, K. S., eds. IPCC Working Group II. Cambridge University Press, Cambridge, UK, 877-912.

Tindale, S. and Hewett, C. (1999), Must the poor pay more? Sustainable development, social justice and environmental taxation. In *Fairness and Futurity: Essays on Environmental Sustainability and Social Justice*, Dobson, A., ed. Oxford University Press, Oxford, UK, 233-248.

Uitto, J. I. (1998), The geography of disaster vulnerability in megacities: a theoretical framework. *Applied Geography* **18**, 7-16.

Woolcock, M. (1998), Social capital and economic development: toward a theoretical synthesis and policy framework. *Theory and Society* **27**, 151-208.

Woolcock, M. and Narayan, D. (2000), Social capital: implications for development theory, research and policy. *World Bank Research Observer* **15**, 225-249.

4. Adaptive Capacity: What Does It Mean in the Context of Natural Hazards?

John Handmer
Centre for Risk and Community Safety
RMIT University, Melbourne, Australia

Many developing countries find that their hopes for development are severely constrained by natural hazards. In some cases, the countries may also be constrained by internal conflict, weak institutions or other social or economic problems. In addition, many scholars argue that the forces of economic globalization are helping to create vulnerable communities throughout the world, but in particular in poorer countries. Clearly, people and communities in these circumstances are likely to find it challenging coping with day-to-day existence, let alone adapt to climate change, which may not even be seen as a relevant issue. The field of natural hazards has much to offer in terms of policy experience and research output. Much effort has gone into examining all aspects of adaptation to hazards—or put more broadly, into examining vulnerability and resilience. Adaptation activity at both the "specific" and "generic" scale has been examined, although research has concentrated on specific measures. The relevance of this work for short and long term adaptation to climate change is examined. It is likely to be high as climate change is most likely to be felt through an increase in weather related hazards. A suggestion is made for situations where neither generic nor specific approaches show promise.

1. Is Climate Change Important?

I would like to start with an assertion: that unfortunately, in terms of action, climate change is not seen as particularly important, nor is it viewed as a high priority in most countries. Even where it is allegedly a high priority, this is often because it serves other purposes—or because it is of interest to the groups developing the policy agendas.

Following the UK floods of autumn 2000 reaction by the media and government officials focused largely on climate change. John Prescott, Deputy British Prime Minister, announced at the UN climate conference in The Hague:

This was a wake-up call that struck home. When people see and experience these ferocious storms, long summer droughts, torrential rains—more extreme and more frequent—they know something is wrong and that climate change now affects them. (BBC News Online, 21/11/2000 "UK floods, a climate alarm call".)

This is a strong statement, but with the blame primarily directed at climate change the pressures to learn and adapt were largely oriented toward global environmental change rather than long standing inadequacies in the government's approach to climate hazard management. It seems likely that the severe flooding of autumn 2000 may be blamed on climate change as a way of shifting responsibility, rather than as a signal to rethink policy. To the extent that this is the case, acknowledging the importance of climate change may actually inhibit adaptation to change by enabling government to escape responsibility for adaptation to climatic hazards. Elsewhere, such as the United States (under George W Bush) and Australia (under John Howard), governments simply deny that climate change is a significant problem.

The paper starts by examining this assertion about the unimportance, or perhaps more accurately the low priority, of climate change for developing countries. After showing that climate change may be very important in certain contexts and arguing that adaptation to climate change would be beneficial and useful—regardless of the extent or direction of change—I examine trends in research on climatic disasters. The research is examined for what it has to say about the "generic" and "specific" approaches to adaptation to climate change. In brief, generic, or macro-scale, adaptive activity includes increasing wealth and education, income distribution, institutions and health care so as to improve countries' ability to cope with climate and other changes. Specific, or micro-scale, adaptive activity is sector-specific, for example planning for sea level rise, increased heat or drought or floodplain management. Before concluding, a third way is suggested for those situations where neither generic nor specific approaches show promise.

2. Coping with Day to Day Uncertainties in Developing Countries

People in developing countries are likely to be much more vulnerable to natural hazards, and are certainly more directly impacted by climatic hazards. The poorer the country, the worst the impacts appear to be. Yet even more than industrialized countries, they have immediate priorities often related to basic survival. The figures concerning health problems, including AIDS, access to potable water, illiteracy and absence of livelihoods, are well known (for example, see the various UN *Human Development Reports*). Some of the general global statistics are even more daunting:

- About half of humanity survives on less than 2 dollars a day—and about half of these exist on less than 1 US dollar a day (UNDP 2001).
- Some 56 countries are currently directly affected by warfare.

Clearly, people and communities in these circumstances are likely to find it challenging coping with their day-to-day existence in ways that most of us cannot begin to imagine, and will be preoccupied with coping with daily uncertainties. How concerned should they be about climate change—which is after all someone else's priority? Their priorities for the future are likely to focus on improving livelihood security, ending war and violence and rebuilding lives. This is not restricted to developing countries; most countries contain substantial groups whose social and economic status forces day-to-day survival priorities.

Of more general importance is simply that political, business and community leaders will almost always have priorities that are a long way from any concern with climate change. This is the case even if their priorities are visionary rather than preoccupied with immediate concerns. Their priorities will concern security, re-election, commercial viability and trade, health, education, infrastructure, environmental management and so on. It is not necessarily that climate issues are seen as unimportant, but they are not as important as other priorities. The relative lack of interest in the International Decade for Natural Disaster Reduction (IDNDR) in terms of profile, resources and political support illustrates the point.

Given the legitimate priorities of many poorer people and countries, and the relative lack of interest among the industrialized nations, it should come as no surprise that few countries are as well adapted as they could be to the present regime of climatic hazards. Adaptation has come about gradually through improvements to building standards, access to insurance and welfare and, importantly in many countries, decreased reliance on agriculture with its acute sensitivity to climate and global markets. This is not so much a criticism as an observation that the best adapted might fall a long way short of the ideal in practice. Good policy on paper needs to go through the process and pitfalls of implementation where it competes continuously for attention, resources and expertise (see Chapters 6, 10 and 11 of this volume).

3. Making Climate Change Important

Despite these serious problems of poverty, war and lack of development affecting so many countries and communities worldwide, climate change is certainly being noticed. It is being noticed through climatic disasters that are drawing attention to change and variability, making climate important and forcing the issue up the political and media priority list.

It is not simply the occurrence of disasters that is important in this context, but rather the apparent increase in their frequency.

The global reinsurer Munich Re (2001) observes that since the 1950s there has been a threefold increase in major natural disasters, an eightfold increase in losses from such events, and a fifteen-fold increase in the losses carried by insurers. The peak year was 1995, at $190 billion or 0.7 percent of global GDP (gross domestic product). This is probably not a reflection of climate change as much as a reflection of the social changes set out below and the global spread of insurance. During the 1990s disasters resulted in a global average each year of 75,250 deaths and 211 million people affected (Walter, 2001).

These figures apply to all disasters other than warfare. From the perspective of our interest in climate change, it is important to note that overwhelmingly these figures refer to climatic hazards, with 90 percent of the deaths from that source. By way of contrast, conflict claimed 2.3 million lives over the same period—or 230,000 per year (Walter, 2001).

4. Disasters and Development

4.1 Disasters and the Developing World

If we are going to tackle climate disasters we need to consider their underlying causes carefully. In doing so it is worth reminding ourselves that disasters interest us only because of their impacts on humanity or things we value. An extreme storm or heatwave is of little concern to us unless it impacts on humanity—apart of course from scientific interest in the natural phenomena. We also need to be aware of where the burden of loss falls, since this may provide additional insights into the causes of disaster and therefore potential avenues for improving the situation.

Most of the human impact of natural disasters is in the developing world. The following figures come from the IFRC (International Federation of Red Cross and Red Crescent Societies) (2001) and draw on their database of 2,557 disasters from 1991 to 2000. They show the dramatic difference between rich and poor countries:

- HDC (highly developed countries): 22.5 deaths per disaster
- MDC (countries with a medium level of development): 145 deaths per disaster (includes China and India)
- LDC (least developed countries): 1,052 deaths per disaster.

It has been conventional wisdom that while developing countries bear the brunt of human losses from natural disasters, developed countries suffered more economically. However, it appears that the poorer the country the greater the impact on the economy and on development progress:

- Honduras (Hurricane Mitch, 1998) = 75 percent of GDP
- Turkey (earthquake in 1999) = 7-9 percent of GDP
- United States (Hurricane Andrew, 1992) = <1 percent of GDP.

Hazards are clearly a vitally important issue for poor countries—even if they are not reflected in budgetary or public sector arrangements. It is to be expected that small countries would be much more affected by a single hazard event than a large country. However, this explains only part of the variation in impact set out above. Many developing countries find that their hopes for development are severely constrained by natural hazards. In some cases, the countries may also be constrained by internal conflict, weak institutions or other social or economic problems.

4.2 Why are "Natural" Disasters Increasing?

There is a widely held perception that there are now more climatic disasters because of the increasing number of climate extremes or severe events. For example, this appears to be the perception in Europe following a series of severe storms, floods and heat waves through the late 1980s and 1990s.

However, as mentioned above this is only part of the story. In fact, the evidence set out above concerning the global distribution of disaster losses suggests that even if there were now fewer climatic extremes we might be seeing increasing disaster losses. This is because social, economic and political factors—factors other than climate—are key to vulnerability and to ability to adapt. The factors that appear to be important explanations of why climatic disasters are increasing are listed here. This list ignores issues surrounding the information on disasters itself. Improved reporting of losses and awareness of all parts of the globe may be contributing to the apparent increase in disaster occurrence.

- *Increasing frequency of climatic extremes*. The evidence is ambiguous, but even if this is the case more climatic extremes do not by themselves result in disaster: climate events must interact with human activity or assets for a disaster to occur.

  But in addition our exposure and vulnerability are increasing because of:
- *Increase in world population*. Any given event affects more people. The evidence for increased population is incontrovertible.
- *Larger increase in poorer areas*. Most population increase is in poor countries that are disproportionately affected by climatic hazards. In addition, many newly occupied areas were previously vacant because they are hazardous. The disproportionate population increases in poor areas, in many cases with a strong increase in areas dependent on agriculture, exacerbate existing wealth inequities and may also increase susceptibility to climatic hazards. The evidence is strong.
- *Dispossession by war or civil strife.* Refugees and those driven into marginal areas are often the most dramatic examples of people vulnerable to the negative affects of natural events—people often completely cut off from their coping mechanisms and support networks. Of the world's (approximately) 191 countries 96 are directly linked to uprooted populations with people being forced to flee in some 60 countries (US Committee for Refugees, 2000).

  The key linkage with climate change here is the increase in vulnerability usually brought on by warfare. Reasons for this increase include an exodus of trained people, an absence of inward investment, disruption of social networks, destruction or abandonment of shelter, redirection of resources from social to military purposes, collapse of trade and commerce, lawlessness and abandonment of subsistence farmlands due to fighting or landmines (see, for example, Levy and Sidel, 2000). In the recent wars in Bougainville and the Solomon Islands, major businesses and the education and health sectors were the first casualties, closely followed by the collapse of livelihoods for many of those driven out of the affected areas. The proliferation of weapons and minefields and the absence of basic health and education can ensure that the effects of war on vulnerability are long lasting. This is particularly the case where the affected area is off the global media radar.

- *Economic and social factors and rapid change*. Chronic corruption, aspects of economic "structural adjustment programs" and the changes accompanying the collapse of communism in Eastern Europe are examples of social factors that undermine people's capacity to cope with hazards. Many countries have serious corruption problems—that is corruption that inhibits development and people improving their lives and prospects. Corruption is by no means restricted to developing countries, but its impacts seem to be far worse in poorer countries.
- *The growth of urbanization*. Much of the hazards literature argues that large contemporary cities are incubators for disasters because of the concentration of people and activities in a confined space and the generation of new hazards (e.g., Mitchell, 1999). However, in many parts of the world—but less so in the world's poorer countries—this situation is balanced by the fact that cities contain massive resources to cope with hazards. In addition, the growth of cities may also be an adaptation against other forms of hazard, including lawlessness and climatic hazards such as drought. The overall situation is unclear, but cities are growing very rapidly and now contain about half of humanity. They are an important focus for future work on adaptation.
- *Economic globalization and gradual environmental degradation*. Again, both these trends may act to undermine the coping abilities of many people. These issues are put together here because of evidence that the factors are often linked. Economic globalization is seen as an unqualified good by almost all political leaders in the industrialized world. The essence of the argument is that through free trade, the whole world will become much richer. However, many poorer countries and those working with poorer sections of society the world over reject this claim. Focusing on distributional issues, they cite as

evidence the growing gulf between nations and between rich and poorer people in all countries. The implication is that the capacity of these poorer groups and countries to adapt to change is declining.

Evidence on these issues is complicated by the strong ideological commitment of both the supporters and detractors. The economies of the major trading nations appear to be benefiting, but much of the world's population is not. Even where per capita incomes in developing countries are rising there is little evidence that this is a direct result of contemporary economic globalization. EU development ministers, the World Bank and other organizations have recently pledged to address the increasing marginalization of least developed countries and the debt issue (Kabissa-Fahamu-Sangonet Newsletter, 2001). One argument is that the process of globalization as managed by the WTO, the World Bank, etc., will not benefit poorer countries because it was not designed to. The original proposal at the Bretton Woods meeting—which established the post-war global financial system—was to create a trade body with much broader powers intended to promote the development of poor areas and restrict corporate power, along the lines of the initial European Economic Community agreement. However, the result was the narrowly focused agencies we have today.

The linkages made with environmental degradation are various; for example, the use of environmentally damaging farming techniques for survival in both rural and urban areas (CEPAL and BID, 2000). The 2002 World Summit on Sustainable Development will focus on "poverty, development and the environment", including trade and the environment, according to South Africa's Minister of Environmental Affairs. He cites the linkages between poverty, environmental degradation, waterborne diseases and the burden on women who bear the burden of daily scavenging for water and energy resources (Moosa, 2001). These are people for whom survival is an effort. Any further environmental degradation, or slight decrease in rainfall, would seriously increase their vulnerability to climatic hazards.

5. Adaptive Capacity Does Not Occur in Isolation

It should be clear that concentrating solely on specific measures to alleviate the impacts of climatic hazards or climate change is unlikely to address the causes of disaster—even though many such measures save lives. Adaptive capacity exists in most communities, but it is unlikely to have been developed solely for climate change. As we can see, promoting adaptive capacity to climate change may require attention to many areas of society, including areas normally seen as well outside the ambit of climate and hazards research and policy. We need to consider what is undermining adaptive capacity or making people more vulnerable. Without doing this we may be attempting to provide a solution to the wrong problem.

If it is thought that capacity to adapt to climate change should be promoted as a special program separate from other areas of society, where should the necessary resources come from? Climate change is invisible, intangible and to many rather esoteric—except of course when it manifests or appears to manifest itself in extreme events that trigger disasters. Why should it receive priority over the many tangible, visible, immediate and urgent needs of the countries concerned, needs like national security, clean water, housing and employment?

Even when the question of priorities is set aside and external funds are available, success is not guaranteed—the measures must be seen as desirable and supported by those at risk or adaptive capacity may not increase. There is an important difference between having measures in place and success. Implementation of policy can be plagued with individual and organizational deviance, unintended consequences, legal problems, conflict with other social goals and expectations and changing political or administrative priorities. The same comments can be made about the actual process of measurement of success (Handmer, 2000).

6. How Can Natural Hazard Research and Policy Experience Support Adaptation?

Natural hazards research brings over 50 years of effort dedicated to working out how to deal best with the problem of climatic and other hazards. And most of this effort has been devoted to what is termed "adaptation" in the climate change literature: much effort has focused on humans rather than on the geophysical phenomena. The field started as a branch of American geography in the human ecology tradition. One of its initial concerns was that despite very substantial investment in engineered structures flood losses were continuing to grow (White et al., 1975). Some might say that it was a reaction against a dominant type of "specific" strategy, that of major flood control works.

Hazards research and policy recommendations in geography initially emphasized the need to consider all options, especially planning approaches, as well as individual choice in location. The evolution of the field's dominant view since the early 1950s is set out in Table 1. More detail on the evolution of thinking by the themes of main emphasis of the research, construction of the problem, and the types of solutions recommended, are set out in Table 2. However, it is important to appreciate that the field is now very diverse, and attempts to capture its main themes will inevitably be inadequate. Please note that here I have followed convention within the hazards research field in which "hazard" refers to the overall problem, in contrast to the emerging international agreement on the use of "hazard" as simply the geophysical phenomenon and "risk" as the whole problem (see, for example, EMA, 2000).

The five stages shown in Table 1 represent dominant thinking at the time indicated. As a general observation, each successive stage shows increased understanding of human and societal behavior, and recognizes that humans have more control over disaster creation than previously acknowledged. Note that different strategies are pursued simultaneously by different groups, and today all five are used in policy.

What does the experience of 50 years of hazards research tell us?

Tables 1 and 2 suggest a trend from specific adaptation toward generic type measures—and this is certainly an important trend. In part this reflects recognition by many hazard researchers of the need to engage with the root causes of vulnerability and disaster. It also shows a realization that specific strategies require appropriate contexts (the "generic" level) for success. The trends in approach to hazard research and management reflect a range of interests and concerns, some of them common to other public policy areas:

- seeking to put hazards research and practice in the mainstream (with sustainable development) and away from a marginal activity
- seeking to deal with causes rather than the symptoms
- looking for long-term solutions

Table 1. Stages in the evolution of dominant thinking in hazards research

1. Individuals are economically rational. Hazards come from nature.
2. Individuals act rationally given the limits of their knowledge and perception. Early 1960s on.
3. Individuals act within constraints set by social and economic structures. Late 1970s on.
4. Social groups organize to share and manage vulnerability to hazards. Hazards are from humans. From the early/mid-1980s.
5. Vulnerability reduction and sustainable development are interlocking aims. From 1990 onward—set out in the IDNDR declaration.

Table 2. Trends in hazards research

From	To
Emphasis:	
Individual decision-making	Community vulnerability
Local	Local—global
Choice	Constraints/social structures
The hazard:	
Hazards as "other"	Hazards are generated by us
Event driven	Situational and less visible creeping hazards included
Solutions:	
Reduce losses	Manage vulnerability or increase resilience
Solutions as separate	Solutions found in organization of society and the development process

- recognizing the limits of individual choice and action
- balancing a long-standing emphasis on specific approaches that appear to have achieved only limited success in many circumstances.

As a general statement we can say that among hazards researchers in the social and policy sciences, adaptation to hazards and sustainable development are now seen as interlocking aims (Mileti, 1999). The 10 year follow-up on the first UN conference on Environment and Development, planned for late 2002 in Johannesburg, has made disaster reduction one of its central themes. In summary, today natural hazards management is largely about being resilient in the face of uncertainty—a description that might well apply to the concept of adaptive capacity for climate change.

Now I would like to briefly examine generic and specific approaches to adaptation in the context of hazards research, taking the generic approach first.

6.1 Generic Approaches in Hazards Research

Much of the recent hazards research and advocacy—that is, literature advocating that efforts be devoted to dealing with hazards—promote the generic approach. Generic approaches generate the wealth, expertise and other conditions needed for implementation of specific strategies such as building regulations. They generally assume effective government and can be expensive. But whatever the difficulties, an overwhelming advantage of generic approaches is that they serve other and often multiple social goals. They are not limited to serving the goal of reduction or management of a specific hazard—important though this might be to those in the field it is rarely a high priority in poor countries. Even when climate is not part of the rationale for generic approaches, they may still reduce the risk by diversifying economies and improving livelihood security by reducing dependence on agriculture or fisheries, and by increasing wealth, so climate change and climatic hazards impact directly on a declining percentage of GDP. Agriculture by definition is highly susceptible to climatic conditions and hazards. In most wealthy countries it is either highly subsidized or compensation and welfare provisions exist—in contrast to poorer countries. Generic measures may also increase resilience to all climatic hazards through improved building, planning and infrastructure, and by increasing or establishing access to resources in a crisis.

A brief overview of some key statements and literature on hazards management in developing countries reinforces the assertion above about the dominance of generic approaches. The development literature on hazards and disasters overwhelmingly supports a generic approach with the emphasis on vulnerability reduction through the development process. For example, the Inter-American Development Bank (CEPAL and BID, 2000) states that "disasters are clearly a development problem". The arrangements for the IDNDR specified that by the end of the 1990s all countries would have disaster reduction plans and activities in place as part of their plans for sustainable development. The follow-up to the IDNDR, the International Strategy for Disaster Reduction (ISDR), makes the link between disasters and sustainable development in its 1999 Declaration of Intent. It emphasizes the need to reduce vulnerability, to safeguard "natural and economic resources, and . . . social wellbeing and livelihoods" (UN/ISDR, 1999). This theme will be taken up at the 2002 World Summit on Sustainable Development where disasters will be on the agenda. In its *World Disasters Report 2001*, the IFRC (International Federation of Red Cross and Red Crescent Societies) takes a similar line: ". . . disasters become ever more frequent, aid dollars and development gains are being washed away. . . . disaster is no longer a brief dip on the curve of development but a danger to the [development] process itself" (IFRC, 2001, p. 1).

The academic literature on the topic generally takes as its reference point the late Fred Cuny's *Disasters and Development* (1983), with its emphasis on broad development issues.

The links between disasters and development, particularly sustainable development, are not confined to developing countries. The theme of the summary report of the U.S. second assessment (Mileti, 1999) of natural hazards emphasizes generic issues and in particular the integration of hazard management with sustainable development.

6.2 Specific Approaches in Hazards Research

To turn to specific approaches—hazards research traditionally has focused on these strategies, although there have always been other voices. They have saved many lives and inestimable amounts of property damage. Some success stories include shelters designed to provide safe havens from sea flooding have saved thousands in Orissa, India (Sparrow, 2001); cyclone warning systems in the Pacific provide timely warnings to poor small island countries; and thousands of flood resistant houses provide protection in Vietnam (Jaquemet, 2001) and many other rich and poor countries. Major levee or dike systems protect the populations of many countries against sea and river flooding. The Netherlands may present the extreme case, where dikes keep the sea out and allow the nation to flourish. However, the situation is not simple because appropriate institutional arrangements and governance are needed for large-scale specific measures.

Opportunities for specific approaches present themselves:

- immediately following a disaster when these measures have high priority
- where there are climate and hazards organizations whose mission includes the use of specific tools
- when local people demand action in the face of perceived risks
- when the institutional context would enable and support low-cost measures such as minor changes to building techniques that can often result in large improvements and locally based warning systems, which are an underused, low-cost measure.

Important constraints are that:

- the measures are often given low priority by governments—and by the people themselves, so those at risk must be involved and must support the measure
- they divert scarce resources from other urgent priorities
- the measures may not be dealing with underlying causes
- occasionally they may make the situation worse.

7. What Do the People Think?

The material set out so far reviews the experience of senior policy makers and the conclusions of the bulk of published researchers. This material is typical in that most of the international literature on development is produced by aid donors, their consultants or academics, mostly from Europe or North America. But what about the opinions of those at risk—the people that all this work and effort is dedicated to helping? They are rarely consulted in a way in which their views might have some influence over the shape of work to reduce vulnerability. (A report by the IIED [2001b] provides some insight into this issue and the views of the urban poor in developing countries.)

Katrina Allen, a PhD student at Middlesex University, undertook a major village level study in the Philippines on vulnerability to flooding (2000, Vulnerability to Flooding in the Philippines, Unpublished paper, Flood Hazard Research Centre, Middlesex University, London). To put it simply, she inquired into the approaches to vulnerability reduction preferred by a major nongovernmental organization (NGO) and government agencies on one hand, and the preferences of those at risk on the other.

The NGO and government agencies adopted the specific approaches of building sea walls and sponsoring counter-disaster training. But those at risk saw the generic or macro issue of livelihood security as key to their resilience. Interestingly, livelihood security was often related to household structure and to having household members sending money gained from employment in a metropolitan area or overseas. Such external income sources continue to flow regardless of local climate disasters and in this way enhance resilience and adaptive capacity. Active kinship networks appear to be more important than community networks in this context.

But is the difference between the NGOs and local people as large as it seems? In general, the two approaches are probably intertwined, with many specific approaches depending on the right generic or macro conditions for their success.

8. Is There a Third Way?

Sometimes neither generic nor specific approaches appear to improve adaptation capacity or local resilience.

In some Pacific island countries disaster management has had limited success with institutional strengthening, generic approaches, the application of geospatial information or more specific measures. Reasons for failure vary across the region. They include violence, weak and ineffective governments, corruption, illiteracy, inappropriate technologies and failure to appreciate cultural issues. Cultural issues mitigating against change include the enormous influence of "tradition", as interpreted by those in positions of power, the importance of village power structures and the general absence of women from formal power structures in many countries even though they may be responsible for the domestic economy including food production and associated specific enterprises such as selling produce. Strong reciprocal obligations among people from the same areas—the *wantok* system of Papua New Guinea, the Solomon Islands and elsewhere—help people cope with climatic and other disasters. However, the rigidity of the system and its imperative of sharing wealth and resources can also inhibit social and economic change as defined by aid donors and the expectations of a capitalist system (see also Chapter 3 of Robertson and Sutherland, 2001).

Under such circumstances a focus on achieving improved resilience by working through governments typically fails regardless of whether generic or specific measures are involved. An appealing solution may be to work on correcting the obvious shortcomings of government. But, this has proved difficult in the circumstances outlined at the beginning of this section. In addition, governments may have different priorities for legitimate reasons—as outlined earlier in this paper.

Is there another way?

A focus on the village or community level may be the only possibility in such cases.

The Australian Landcare program provides an interesting example of this approach. I am not suggesting that the program could be directly transferred to other countries—simply that it is an interesting approach. The official websites (e.g., www.landcareaustralia.com.au) state that Landcare is a high profile movement or network of people in Australia who are committed to sustainable management and use of natural resources. Landcare has been about social change, a change in the ideas and practices of people who work on and care about the land. The driving logic is that the other approaches tried by government did not deliver the desired sustainable outcomes. Today there are almost 5000 Landcare groups working in communities throughout Australia. Groups consist of local landowners and others with an interest in the land or a local problem. The groups are generally supported by facilitators and coordinators funded by government and may apply for project grants. Many groups attract commercial sponsorship. Despite its apparent success, Landcare has attracted intense debate and criticism (see, for example, Curtis et al., 2000).

Whatever is said about the national framework and about the degree of inclusiveness, governments abrogating their responsibilities and tangible achievements on the ground, the Landcare program is also about a process for local people to identify their problems and develop locally appropriate solutions. The program recognizes that valuable local resources and expertise exist which with support can deal with many local problems. It is the recognition of the value of local resources and process elements, in addition to the approach's ability to harness informal networks, that may be relevant for some developing countries. The Landcare program is coordinated and supported by government, but other groups could be substituted for government where appropriate.

Where adaptation is concerned, a community or village based approach is where those at risk are helped through a deliberative process to decide what (adaptation) actions are appropriate. This help would generally take the form of assistance with the decision process including any necessary technical advice. The local groups would then receive support so that they can implement their decisions. Ultimately, the process would build local support and capacity for adaptation. Possibilities include local warning systems and emergency preparedness, improved building practices and livelihood strategies. Another approach, based on similar principles of local organization and action, is set out in a briefing paper by the International Institute for Environment and Development (IIED, 2001a).

9. Conclusions

Disasters frequently demonstrate the importance of climate events. This is the case particularly for the least developed countries, which bear the brunt of human losses and are also most likely to see economic progress reversed by climatic disasters. Governments and communities around the world face many challenges of which climate change is only one, and rarely one that rates a high priority in a context of a daily struggle for survival by half the world's population, and the imperatives of trade, security, health and so on.

It is hardly surprising therefore that most countries could be much better adapted to the *existing* natural hazards regime—and fall far short of dealing seriously with future climate change. There are many individual examples where future climate has been incorporated into design and policy, but these are of an ad hoc nature and rarely extend to poorer countries.

The low priority and status of climate change in policy means that most specific or micro adaptation is likely to be an extension of existing natural hazard management. It is unlikely that new programs for climate change would be implemented successfully unless they ran in parallel with other high priority issues. But generic programs with their multiple aims might be possible.

The research and policy experience of the natural hazards field has much to offer the search for adaptation to climate change as the problems are very similar. Research suggestions are set out in Table 3.

Much of the hazards and development literature and development and humanitarian aid groups now emphasize approaches to hazards management

Table 3. Research suggestions

- Spend time defining the problem—to avoid providing the right answer to the wrong problem. If the real problem is weak governance, war or corruption, then that needs to be considered and addressed—or a way found to avoid these impediments.
- Ways of ensuring synergism between specific and generic approaches are needed.
- Related to both of the above issues is the issue of how to improve the management capacity of households, communities and governments so that they are better able to cope with climatic hazards.
- Climate change concerns need to be mainstreamed, e.g., with sustainable development issues.
- Indicators of adaptation and of resilience need to be developed.
- What to do where both generic and specific approaches fail? Should a village level approach be developed?
- Processes are needed to ensure that the informal, the traditional and local are part of adaptation strategies. (I appreciate that "traditional" is always the subject of debate and often invented for commercial or political reasons.)
- The role of rapidly expanding cities in adaptation should be examined and opportunities harnessed.
- What should the role of the private sector be?

which focus on causes rather than symptomatic treatment of the problem. However, there appears to be a near symbiotic relationship between these generic development strategies and specific approaches aimed at alleviating identified hazards, suggesting that both are needed to reduce vulnerability to climatic hazards. Despite their costs and other problems, generic approaches can be seen as "no regrets" strategies. In other words, they are generally worth putting in place even if their effectiveness against hazards is uncertain provided they have other potential benefits. These benefits normally take the form of improvements in for example, health, education, housing quality or livelihood security.

However, sometimes neither generic nor specific approaches seem effective, often because of the inability to create an adequate institutional framework. In such circumstances, a "village" level approach should be developed because this may be the only effective way forward.

An important outstanding issue concerns the assessment of progress in adaptation. We need to be able to say with some authority whether or not adaptation is occurring, and how effective it is. The basis of these assessments needs to be clear. The best way to proceed may be through the development of indicators of adaptation.

Acknowledgements

My appreciation goes to Richard Klein and Joel Smith for organizing the Potsdam Workshop and for inviting me to participate. Thanks also to Cassia Read for research assistance during the preparation of this paper.

References

CEPAL and BID (Economic Commission for Latin America and the Caribbean and Inter-American Development Bank) (2000), *A Matter of Development: How to Reduce Vulnerability in the Face of Natural Disasters.* CEPAL, Mexico.

Cuny, F. C. (1983), *Disasters and Development.* Oxford University Press, New York.

Curtis, A., van Nouhuys, M., Robinson, W. and Mackay, J. (2000), Exploring landcare effectiveness using organizational theory. *Australian Geographer* **31**, 349-366.

EMA (2000), *Emergency Risk Management—Applications Guide.* Emergency Management in Australia, Canberra.

Handmer, J. (2000), Are flood warnings futile? *Australasian Journal of Disaster and Trauma Studies.* http://www.massey.ac.nz/~trauma/ (Journal issue: 2000-2).

IFRC (2001), *World Disasters Report 2001.* International Federation of Red Cross and Red Crescent Societies, Geneva.

IIED (2001a), Deep Democracy: Transforming Opportunities for the Urban Poor. Environment and Urbanisation Brief 4. International Institute for Environment and Development, London.

IIED (2001b), Rethinking Aid to Urban Poverty Reduction: Lessons for Donors. Environment and Urbanisation Brief 3. International Institute for Environment and Development, London.

Jaquemet, I. (2001), Post-flood recovery in Viet Nam. Chapter 5 in *World Disasters Report 2001.* International Federation of Red Cross and Red Crescent Societies, Geneva.

Kabissa-Fahamu-Sangonet Newsletter (2001), PAMBAZUKA NEWS. 43, 22/11/01. http://www.kabissa.org

Levy, B. S. and Sidel, V. W. (2000), *War and Public Health.* American Public Health Association, Washington, DC.

Mileti, D. (1999), *Disasters by Design: A Reassessment of Natural Hazards in the United States.* Joseph Henry Press, Washington, DC.

Mitchell, J. K., ed. (2000), *Crucibles of Hazards: Megacities and Disasters in Transition*. UN University Press, Tokyo.

Moosa, M. V. (2001), Johannesburg 2002 must be about people, planet and prosperity. In *Environment Matters at the World Bank.* The World Bank, Washington, DC, 4-5.

Munich Re (2001), www.munichre.com/index.html.

Robertson, R. and Sutherland, W. (2001), *Government by the Gun: The Unfinished Business of Fiji's 2000 Coup.* Pluto Press, Annandale, Australia.

Sparrow J. (2001), Relief, recovery and root causes. Chapter 1 in *World Disasters Report 2001*. International Federation of Red Cross and Red Crescent Societies, Geneva.

UNDP (2001), Human development: past, present and future. Chapter 1 in *Human Development Report 2001*. UN Development Programme, New York.

UN/ISDR (1999), International Strategy for Disaster Reduction, 1998 Declaration of Intent. http://www.unisdr/org/isdrdecla.htm.

US Committee for Refugees (2000), *World Refugee Survey 2000*. US Committee for Refugees, Washington, DC.

Walter, J. (2001), Disaster data: key trends and statistics. Chapter 8 in *IFRC World Disasters Report 2001*. International Federation of Red Cross and Red Crescent Societies. www.ifrc.org/publicat/wdr2001/chapter8.asp

White G. F. (1975), *Flood Hazard in the United States: A Research Assessment.* Institute of Behavioral Science, University of Colorado, Boulder, USA.

5. Lessons from Famine Early Warning and Food Security for Understanding Adaptation to Climate Change: Toward a Vulnerability/Adaptation Science?

Thomas E. Downing

Stockholm Environment Institute, Oxford Office

10b Littlegate Street, Oxford, United Kingdom

Famine early warning systems, and food security research more generally, were developed to address a specific problem—supporting local livelihoods and national economies to prevent widespread adverse effects. This experience, over two decades, provides insight into some of the challenges, and potentially solutions, in adaptation to climate change. Progress toward a vulnerability/adaptation science has been significant, but a dynamic understanding of vulnerability and adaptation pathways will require a change in paradigm.

1. Introduction

Most research on climate change impacts in the 1980s and 1990s adopted one of two approaches. Regional (and some global) studies tended to be sectoral assessments, often linking climate scenarios with biophysical impacts or studying responses of individual "exposure units" to selected climatic threats (and less often opportunities). Global integrated assessments, and some regional studies, that looked across sectors tended to do so based on abstract mechanisms (e.g., markets to mediate resource allocation) or with limited and constraining interactions between sectors (e.g., some simulations resulted in continental deforestation in Africa because land was required for achieving food self-sufficiency). Few if any significant studies captured the dynamism of local vulnerable groups, current exposure to risks and stresses, processes and shocks at greater scales and the synchronous responses to climate variations that define future vulnerability to climate change and the requirements to build adaptive capacity.

In contrast, research and practice on food security and famine early warning in those decades sought more sophisticated means to integrate the local and global, to understand exposure across a range of shocks and to decipher the climate-food chain of causation against a background of the human ecology of production, exchange entitlements and political economy, to refer to a common construction of the social vulnerability of hunger.

This paper reflects on that progression of learning to adapt to risks in the sphere of food security. Earlier versions (Downing, 2002, and forthcoming) reviewed the extent of the risk (e.g., climate change impacts), notions of vulnerability, and prospects for adaptation (see also Downing et al., 2001). Here, I take as the starting point eight "lessons" learned in the practice of food security and famine early warning. Within each lesson, I suggest substantive themes that link vulnerability and adaptation (using agriculture and food security as examples) and provide samples of methods that could bring insight into processes of adapting to climate change.

Food security assessment and famine early warning are mature areas of research showing considerable progress over the past decade. This is itself a measure of increasing adaptive capacity with respect to climate change. Moreover, it provides a rich experience to contribute to climate change adaptation methods and capacity building. The lessons learned are distilled from the experience of such efforts as the UN's Food Insecurity and Vulnerability Information and Mapping Systems (www.fivms.org), World Food Programme's vulnerability mapping (www.wfp.org/operations/vamu), Save the Children Fund's RiskMap and household food economy approach (http://193.129.255.93/foodsecurity), assessment processes by the US Famine Early Warning System (www.fews.org) and CARE's livelihood security framework (www.careinternational.org.uk/reports/livelihoods.html). However, these are personal observations.

2. Beyond Hoovering: Conceptualizing Vulnerability

The early models of famine early warning systems adopted the approach of gathering all possible (or measurable) indicators, adding them up and deducing vulnerability. This was termed the hoovering approach—

vulnerability was the weight of the bag after vacuuming up everything in sight (hoovering is the British word for vacuuming). The dangers, of course, in the indiscriminate use of indicators are that many indicators are correlated with each other, some have weak or delayed connections to emerging vulnerability and indicators that are difficult to measure may be ignored.

Subsequently, analysts emphasized logical, conceptual approaches to vulnerability and used conceptual models to choose relevant indicators. Many agencies appear to have converged around a broad framing of vulnerability (see Figure 1 for CARE's livelihood approach). An explicit framework provides a common language, the basis for progressive discussion of appropriate interventions, and a means to communicate detailed findings to relevant policy audiences.

Without repeating the many vulnerability frameworks and definitions in existence, Figure 2 contrasts their orientations. Top-down approaches tend to be based on some theoretical basis while bottom-up approaches are more closely connected to local situations and data. The match between the method and understanding of vulnerability processes varies from a reliance on indicators (usually ad hoc), use of formal conceptual frameworks and practical applications (based on some empiricism).

In contrast to practice in food security, the climate change community has not reached consensus on the conceptualization of vulnerability. The Intergovernmental Panel on Climate Change (IPCC) created a distinct definition of vulnerability, essentially the residual impacts of climate change after adaptation measures have been implemented. This approach confines vulnerability to long-term changes driven by climate change. It is difficult to operationalize since it mixes present vulnerability, long-term scenarios of climate change and multilevel processes of adaptation.

3. Baseline Exposure: Who, to What, When?

Monitoring emerging famine conditions requires an appropriate baseline of vulnerability. Much early discussion revolved around whether vulnerability was socioeconomic or geographic (who?), structural or episodic (to what?), and long term or short term (when?).

Vulnerability is more characteristically about people rather than places. That is, vulnerability is a social phenomenon relevant to particular social groupings, whether demographic (elderly, young), economic (livelihoods,

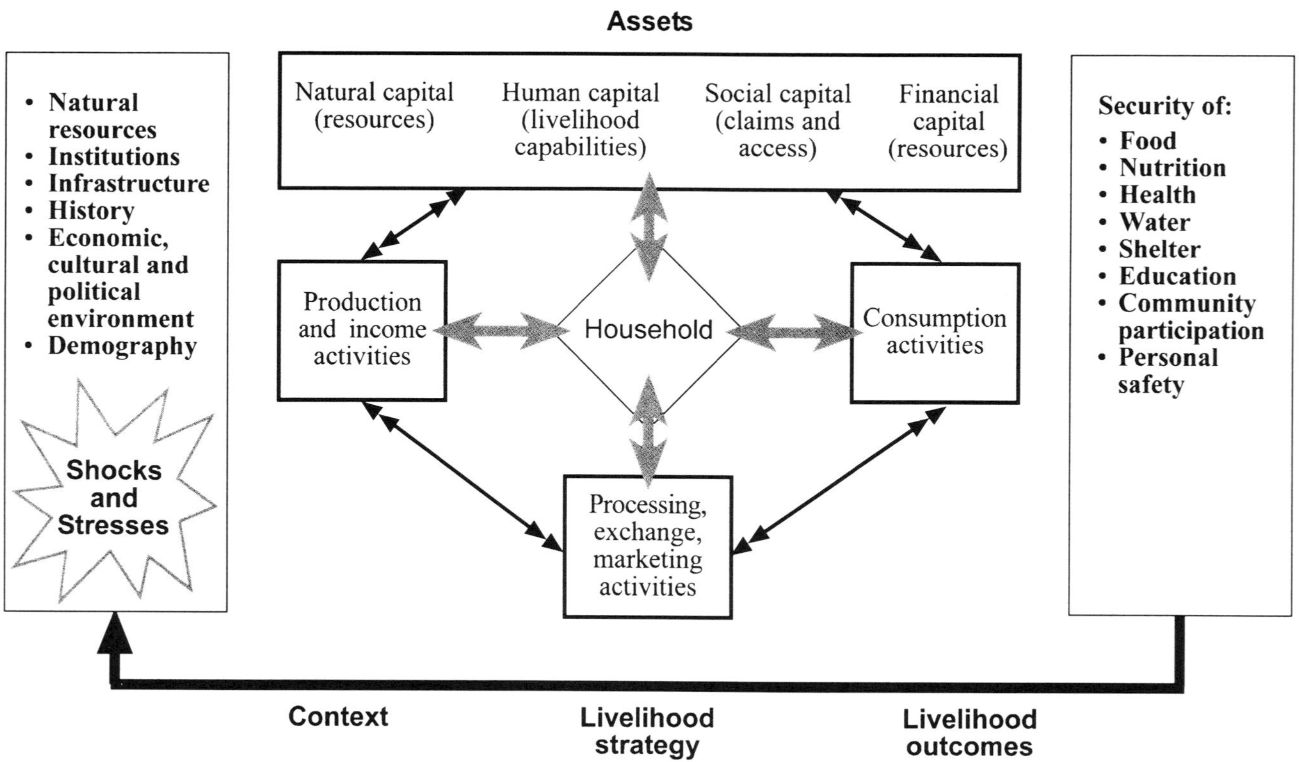

Figure 1. CARE's livelihood security framework

	Indicator-led	**Conceptual**	**Practice-based**
Top-down	Indiscriminate indicators: a 'round-up' of indicators, all of which are considered important, with no *a priori* structure (e.g., Environmental Sustainability Index)	Based on theory; useful to select vulnerable groups and relevant indicators (CARE livelihood model)	Based on convention, useful to highlight specific issues, often for policy purposes
Bottom-up	Wide range of data collected, analysis using clustering techniques (e.g., classified regression techniques)	Participatory identification of vulnerable groups and processes, selection of key indicators (e.g., profiles)	Empirical vulnerability model developed to predict an observed outcome (e.g., under-nutrition)

Figure 2. Vulnerability approaches. Top-down and bottom-up refer to source of the framework and relationship to local data. The columns reflect the degree to which the framework is conceptual with a theoretical underpinning, or based on either an ad-hoc selection of indicators or conventional practice. Best practice would combine many elements of these examples. See Stephen and Downing (2001).

entitlements) or political (marginalized). It is people who are exposed (ultimately) to climatic extremes, and to longer term climate change.

The structural versus episodic debate continues, at least in terms of the interrelationship between episodes that trigger vulnerability and the ability to recover (e.g., resilience as a condition of adaptive capacity).

The time period of baseline vulnerability is generally taken to be the recent present—that is, the realization of historical trends and events that are relevant to the current snapshot of vulnerability.

Understanding climate change adaptation requires construction of a baseline. The current trend in research is to construct (artificial) scenarios of future socioeconomic conditions and compare potential impacts against those futures. While this approach has some resonance in the comparative statics, for example of a drought impact, it does not capture the dynamics of vulnerability and adaptation processes (see below); neither does it provide sufficient insight into the nuances of present vulnerability.

Exposure to food insecurity, for both climatic disasters and longer term climate change, provides examples of the issues of defining the baseline for both vulnerability and adaptation. The baseline is strongly influenced by scale. At the local scale, the actors are the vulnerable populations and their livelihoods, often centered on the household and its relations with its community. The dominant framing of vulnerability at this scale is generally the human ecology of production—resources available and used, decision-making regarding land use and cultivation, gender and labor issues. The effects of climate change can be modeled at the site scale, taking account of direct effects of CO_2 enrichment, warmer conditions, altered precipitation regimes and associated effects on crop and livestock pests and diseases. Evaluation of detailed studies at this scale suggests that the sensitivity of individual cultivars needs to be tempered by potential agronomic adjustments (e.g., changed planting dates), adoption of existing cultivars (e.g., ones adapted to shorter growing seasons) and prospects for technological change (e.g., gene manipulation to enhance the CO_2 effects on water use efficiency to overcome precipitation changes).

The local is fixed in a national/regional scale, the economic and policy institutions that govern resources. At this scale many studies have adopted techniques in spatial crop climate modeling, either using reduced form models or linking site and regional approaches. Such assessments add to the site adjustments with spatial adaptations, such as shifting the cropping area. However, the main issues for food security at the national/regional scale are the operation of the market and the political economy of resource allocation. Economic exchanges may be captured in economic models (regarding land use allocation, farm decision-making and national input-output accounting). Few if any model assessments account for the processes of marginalization at this scale—the nuances of expanded exchange entitlements and the political economy of agricultural and rural development.

At the global scale, the focus shifts to concerns for planetary food self-sufficiency (at a facile level) and the distribution of agricultural between countries (see Box 1 for a good example). Some of the most interesting, and frustrating, research has been at the global scale. Crop climate modeling tends to be overly simple, and often at odds with the more detailed understanding cited at the local level. Economic models are generally sectoral, with many critical variables exogenous to the assessment

Box 1. Exposure to food insecurity using climate change scenarios

Three conclusions regarding adaptation to climate change are illustrated by the IIASA assessment (Fisher et al., 2001, 2002). First, at the global level, total food production is not a real threat. However, the distribution of food will require greater trade to balance out the winners and losers. Current cereal production among the winners is 568 million tons, against production in those countries that lose agro-ecological potential of 475 million tons. Among the winners, in the climate scenario shown in Figure 3, potential increases by over 20 percent (128 million tons), while the losers suffer losses of 7 percent (33 million tons). Such impacts would require considerable increases in trade, particularly to cope with increases in demand related to population and economic growth in developing countries.

Second, the regional impacts are quite diverse and not yet predictable from the range of climate scenarios. The effects of climate change will be patchy, often with regions of risk and benefit within countries. For example, higher resolution modeling in Kenya shows the highlands may gain in terms of potential cereal production while desiccation and drought become worse problems in the lowlands. Regional investment may depend on the spatial patterns of expected changes. Agricultural capital is likely to be sensitive to medium-term prospects. Already planners need to identify regions where the return is sufficient for increased investment.

Third, at the local level, changes in livelihoods and food security are expected. The 10 countries with high rates of malnourishment today might see their national food balance deficit increase, from 5 to 7 million tons per year. Perhaps equally worrying are the 11 countries with lower rates of malnutrition that might see their food gap increase by an order of magnitude (from 1 to 13 million tons). On the other hand, some 38 countries might see gains in production that more than offset current food gaps. The focus on food security, even at this national level, brings to the fore adaptation options related to rural development options such as health systems, employment (especially off-farm) and agro-technology.

(e.g., economic growth and technological change, which may be justified where agriculture is a small proportion of the economy, but raises doubts about the ability to provide insight into regional threats and effects on the most vulnerable).

As an example of crop-climate sensitivity, Figure 3 shows the change in potential cereal production for one climate scenario. The tally of winners and losers reflects that large areas of northern middle-latitudes become suitable for production, especially in Canada, Russia and China. In contrast, tropical countries are more likely to suffer adverse consequences. Gains in cereal potential are shown for 24 developed countries and 68 developing countries, but the gains among developed countries are much the larger. Losses are modest for the 17 developed countries, less than half of the losses shown for 49 developing countries.

While "stakeholder participation" has become a mantra in integrated assessment, few studies begin with an analysis of stakeholders—the who in vulnerability and adaptation decision-making. Adapting to climate change is not automatic or only autonomous. The motivations, constraints and domains of authority of decision makers involved in shaping policy, implementing decisions and coping with the consequences of changes in resources and hazards must be considered. The principal stakeholders range from vulnerable consumers to international organizations charged with research and relief. Stakeholders will suffer the consequences of climate change to varying degrees and have primary concern for different types of adaptations. This is likely to influence their involvement in planning and implementing adaptive responses.

4. An Integrated Measure of Exposure

Vulnerability integrates exposure at the level of vulnerable groups; therefore vulnerability must span sectors and sources of shocks. The concern in famine early warnings is with an outcome—food security. The sources of insecurity may be differentiated, but the key focus is preventing loss of life, assets and livelihoods regardless of the source of stress. Sen's (1981) work on entitlements marked the end of the single stress-impact era of food security work (that is, the linear construction of famine as drought → crop failure → production decline → famine).

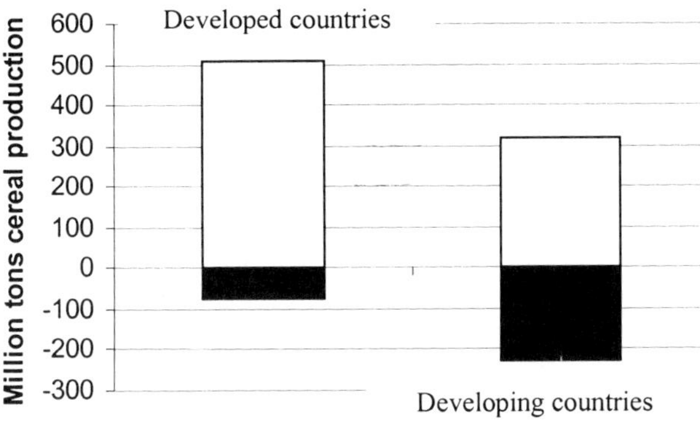

Figure 3. Impact of climate change on developed and developing countries. This assessment is based on the FAO/IIASA agro-ecological zone model that calculates crop suitability based on land quality (e.g., altitude, topography, soils) and climatic suitability (temperature and precipitation ranges). The climate scenario is Max Planck Institute for Meteorology ECHAM4 for 2080, using present cropping systems including multiple rain-fed cereals. The bars show the gains and losses in production, for countries grouped as developed or developing. Source: Fisher et al. (2001).

Similarly, sustainable livelihoods approaches rely on a multidimensional understanding of vulnerability. In the CARE example (Figure 1), this is expressed as four "capitals": natural, human, social and financial (Table 1).

In contrast, most climate impact studies start with the assumption that climate stresses are foremost. The most common framework remains: climate change scenario → impact model → climate change impact. Modest variations to this linear view include the use of analogue events (e.g., a signal drought) and changes in variability (and therefore risk) in addition to changes in climatic means.

However, the more recent focus on adaptive capacity (as somewhat distinct from specific adaptation measures) has brought back into climate policy notions of integration with development—variously labeled mainstreaming, enabling environment or capacity building. In any case, adaptive capacity is seen as "beyond wealth", to include technology, institutions, social networks, etc.—the kinds of processes underlying notions of sustainable livelihoods. There is still some way to go before the inventories of adaptive options can be integrated with the broader assessments of adaptive capacity. And there is no guarantee that capacity at

Table 1. Sources of risk to household livelihood security

	Types of risk				
		Social			
Sources of livelihood	Environmental	State	Community	Economic	Conflict
Human capital (labor power, education, health)	Disease epidemics (malaria, cholera, dysentery) due to poor sanitary conditions, AIDS	Declining public health expenditures, user charges, declining education expenditures	Breakdown in community support of social services	Privatization of social services, reduction in labor opportunities	Conflict destroys social infrastructure, mobility restrictions
Financial and natural capital (productive resources—land, machinery, tools, animals, housing, trees, wells, etc.—liquid capital resources—jewelry, granaries, small animals, savings)	Drought, flooding, land degradation, pests, animal disease	Land confiscation, no secure tenure rights, taxes, employment policies	Appropriation and loss of common property resources, increased theft	Price shocks, rapid inflation, food shortages	Conflict leads to loss of land, assets, and theft
Social capital (claims, kinship networks, safety-nets, common property)	Recurring environmental shocks breakdown ability to reciprocate. Morbidity and mortality affect social capital	Reduction in safety net support (school feeding, supplementary feeding, FFW, etc.)	Breakdown of labor reciprocity, breakdown of sharing mechanisms, stricter loan requirements, lack of social cohesion	Shift to institutional forms of trust, stricter loan collateral requirements, migration for employment	Communities displaced by war, theft leads to breakdown in trust

Table 1. Sources of risk to household livelihood security (cont.)

	Types of risk				
		Social			
Sources of livelihood	Environmental	State	Community	Economic	Conflict
Sources of income (productive activities, process and exchange activities, other sources of employment, seasonal migration)	Seasonal climatic fluctuations affecting employment opportunities, drought, flooding, pests, animal disease, morbidity and mortality of income earners	Employment policies, declining subsidies or inputs, poor investment in infrastructure, taxes		Unemployment, falling real wages, price shocks	Marketing channels disrupted by war

Source: CARE, from a workshop of the Sphere Project (2001).

one level transfers to effective resource management at another level. Indeed, we should expect conflicts over the policy lead in climate change to preclude effective adaptation in some regions and sectors.

One way to capture two of the dimensions of vulnerability—current development and exposure to future climate change—is to use indices to cluster vulnerable situations. Figure 4 illustrates this approach, charting development opportunities and challenges according to current adaptive capacity and vulnerability to future climate change impacts. The four quadrants represent qualitatively different prospects. The most serious concerns are for the upper left quadrant—where present adaptive capacity is low and climate change impacts are likely to be high. For example, livelihoods in the coastal zone of Bangladesh are almost certain to fall in this category of vulnerable communities. Further refinement in climate predictions is not necessary to assign a high priority to increasing adaptive capacity. Where impacts could be high but adaptive capacity is also high—coastal resorts in Antigua, for example—development assistance could lead to new opportunities and a sustainable strategy to cope with increased risks (the upper right quadrant).

Where impacts are low, further information on the nature of climate change may be warranted, although a precautionary approach to situations of low capacity may be suitable to be prepared for surprises (such as a rapid increase in damaging extreme events), as in the lower left quadrant. Where impacts are expected to be relatively low and capacity high, there should be little impediment to achieving sustainable livelihoods.

The first step in operationalizing the clustering approach, as a pilot demonstration, was to devise an index of aggregate climate change impacts. In the increasing number of global climate change impact studies, few include more than one sector at the country level. For illustrative purposes, an earlier study commissioned by the EU ExternE project (with support from the International Energy Agency's Greenhouse Gas R&D Programme) provides the requisite scale and coverage (Downing et al., 1997). Impacts are estimated by linking a climate change scenario (in this case only one—the IS92a linked to the early climate change scenario produced by the Goddard Institute for Space Studies) to a first-order biophysical impact model (such as a simple water balance). The change in the impact model is then valued according to various econometric techniques (none of which should be taken as wholly reliable).

	Adaptive capacity	
Impacts	**Low**	**High**
High	Vulnerable communities	Development opportunities
Low	Residual risks	Sustainability

Figure 4. Climate change adaptation matrix. The quadrants represent artificial boundaries of our knowledge of anticipated impacts of climate change and capacity of livelihoods or regions to adapt to climate change impacts.

The sectors included in the impact models at the country level are agriculture, biodiversity, water, cooling energy, heating energy and sea level rise. To compile an aggregate index, the country-level estimates of climate change damages for each sector were multiplied by a weighting factor (which values agriculture, water and sea level rise impacts higher than energy impacts), and then the sectoral costs were summed and normalized. To illustrate issues of food security, the weightings were 40 percent for agriculture, 25 percent each for water and sea level rise and 10 percent for biodiversity (the other sectors were not used; i.e., their weights were 0).

The quintiles of the impact index were determined. In the following clusters, the first two impact quints (40 percent of the countries) were considered high potential impacts.

The index for adaptation is much simpler—this experiment uses the Human Dimensions Index as an indicator of relative adaptive capacity (UNDP, 1999). The HDI is itself an aggregate of indicators on life expectancy, gross domestic product (GDP) per capita, and education. The first three quints of the HDI were assumed to be countries with relatively low adaptive capacity.

The countries were grouped (Figure 5) according to their estimated impacts and adaptive capacity, using the matrix shown in Figure 4. The most vulnerable countries (high impacts and low adaptive capacity) contain over 3 billion people, with an average per capita GDP of less than US$4,000 (Table 2). The opposite situation—where sustainable livelihoods are not likely to be threatened by climate change (low impacts/high capacity) contain less than half a billion people, and their average GDP is about US$12,000.

Many wealthy countries can also expect high climate change impacts in terms of the global cost of climate change—nearly 750 million people live in these countries. The most uncertain class is where the aggregate index of impacts is relatively low, but adaptive capacity is also low—in these countries of residual risk climate change could contain some surprise or be a serious threat to some livelihoods within those countries. They comprise about the same number of people as the sustainable livelihoods cluster, but with about two-fifths of the GDP per capita.

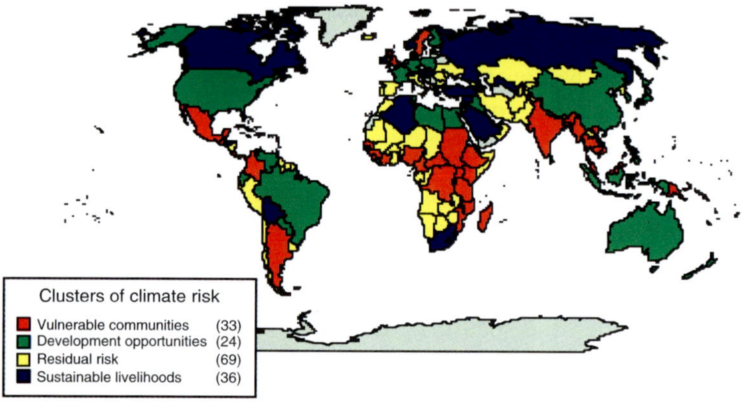

Figure 5. Clusters of climate risk, according to estimated impacts and adaptive capacity

Table 2. Clusters of climate risk

Population, 1000s				Gross Domestic Product, per capita (US dollars, 1995)			
	Adaptive capacity				Adaptive capacity		
Impacts	Low	High	Total	Impacts	Low	High	Average
High	3,409	743	4,152	High	3,662	19,614	6,516
Low	703	316	919	Low	5,003	12,077	7,195
Total	4,112	1,059	5,170	Average	3,891	17,367	6,650

5. A Relative Assessment

Vulnerability is not an external characteristic that can be universally described and observed. Rather it is a way of addressing complex problems and deciding on a course of action—whether mobilizing aid to prevent famine or mapping underlying food insecurity to target food aid programs. The analyst must choose indicators, aggregation models and relevant thresholds for action. For example, a common approach is to simply add all of the selected indicators together—the hoovering approach. Yet, stakeholders may have very different mental maps of vulnerability—some will key on just a few indicators (poverty, for example), others will seek relationships between the triggering event and its consequences (as in the standardized precipitation deficit), and yet others will be concerned with social networks, governance and institutional capacity (after all, what is the point of know who is vulnerability if it is unrealistic to do much about it). Recognition that vulnerability assessment is not directly observable should lead to reflective and interactive approaches. This is still a challenge to practitioners, although developments in interactive geographic information systems are promising and pilot studies are becoming increasingly sophisticated (see www.povertymap.net).

The search for significant thresholds is common in climate impact studies, whether as part of global integrated assessment models (see the Questions project at the Potsdam Institute for Climate Impacts Research, www.pik-potsdam.org) or in defining coping ranges (see Jones and Page, 2001). In one sense this parallels the concepts of relative vulnerability and the emphasis on a defined exposure (the who in vulnerability assessment). However, thresholds are not immutable boundaries—they are at best heuristics that reflect the current understanding of the sensitivity of complex systems to shocks. This has been widely accepted in famine early warning and disaster management—the definition of a crisis is situated within the economic and political context simultaneously at local and global scales.

6. Targeting: Scale and Specificity

Early food security assessments that focused on food deficits may have missed the impacts of economic shocks—the motivation for Sen's (1981) entitlement system. Preventing famine requires targeting the most vulnerable groups, at least to use the existing resources efficiently. To support livelihoods (rather than send a lot of food aid), increasingly sophisticated food security assessments are necessary—for instance taking on local pockets of deprivation, understanding the role of local markets within regional economies and anticipating changes in remittances and household expenditures. For example, the World Food Programme approach involves a national screening of indicators (e.g., changes in vegetation quality), then selection of key areas for local livelihood assessments.

Scale and specificity are inherent in climate change adaptation. The common catalogue of adaptive responses is shown in Box 2. The measures and strategies are roughly ordered according to the scale of implementation, from local farm management and livelihood security to national and international action. Many of the actions can be called no regrets—either they have widespread (ancillary) current benefits or their costs are small relative to short-term benefits as well as long-term adaptation to climate change. However, some are costly and require significant infrastructure and development funding. Other measures can lead to capacity building, or require a strategic framework to be implemented.

7. Capturing Dynamic Pathways and Multilevel Processes

Vulnerability changes quickly, the product of changes at several scales. Most assessments focus on current situations, in order to recover from a disaster or anticipate a crisis. The baseline of recent vulnerability is likely to vary from a few years (e.g., trends in household income) to a few months (e.g., crop failure) or even the past few days (as in situations of armed conflict). While the dynamism of vulnerability is recognized, there has been relatively little progress in capturing evolving pathways or projecting vulnerability into the future. A starting place has been to focus on indicators that reflect changes rather than states (Anand Patwardhan, presentation to the Workshop on Climate Change Vulnerability and Adaptation, Trieste, Italy, June 2002).

Representing climate impacts and adaptation as dynamic pathways is also a challenge for climate change studies. It may turn out that the fixation on scenarios—as static snapshots of futures that have little connection to the present—has impeded more creative development of dynamic representations of environmental change and social learning. I believe three elements are required for this method to be fully developed (and most of these are under way to some extent):

Box 2. Adaptation measures and strategies for agriculture

The most certain aspect of climate change is increased CO_2 concentrations. Efforts to enhance the positive CO_2 responses in new cultivars may be worth the investment in plant breeding and agricultural technology, irrespective of changes in moisture availability. This research and development program would be undertaken by national or international research centers—as recommended by the May 2002 Expert Workshop on Climate Change and Agriculture in South Asia sponsored by the Government of India, UN Environment Programme and Consultative Group for International Agricultural Research.

Inexpensive adaptation includes establishing strategic food reserves to buffer potential increases in the variation of local and national production. This is suitable only at the national or international level. Connections to local grain banks are required, but increasing local reserves because of climate change may not be warranted at present.

Measure/strategy	No regret	Infrastructure	Strategic capacity building
Reduce production subsidies	✓		
Drought preparedness, warning and management	✓		✓
Avoid monoculture, diversify crops	✓		
Conserve soil moisture and nutrients	✓		
Diversity income, off-farm employment	✓		✓
Crop, farm and income insurance			✓
More R&D on heat and drought resistant varieties, maximize CO_2 effects			✓
Tailor land use planning to consider potential climate change			✓
Reduce runoff, improve water uptake, reduce wind erosion	✓	✓	
Increase irrigation efficiency, prevent salinizations	✓	✓	
Upgrade food storage and distribution systems	✓	✓	
Liberalize agricultural trade	✓		✓
Increase food aid			✓
Economic safety nets, welfare systems			✓
Change dietary preferences			✓

Adapted from Downing et al., 1997.

Box 2. Adaptation measures and strategies for agriculture (cont.)

Protection against present and future extreme events should be a priority. Drought early warning and preparedness is urgent, building on the considerable improvements that are already under way in many regions of Africa. Making better use of climate predictions is a key aspect of this adaptive strategy, but it is not a facile connection between coping with variability and coping with long term change.

Market structures often support crops with a high level of risk and fail to support markets for drought-tolerant crops (see the Mexico case study in Chapter 11 in this volume). On the other hand, restrictions on trade may increase the volatility of local markets and food supplies in response to climatic variations. The most effective response to changing climatic risks is not easy to predict, probably cannot be generalized across regions and almost certainly is not as simple as ensuring that markets are "efficiently allocating resources".

- The adaptation paradigm needs to be connected more intimately with the climate change/impact work. Some adaptation specialists have argued that the marginal value of a new climate scenario is near zero (and some even argue it is negative) since adaptation processes are relatively short term and local, and require probabilistic projections. Yet, adaptation to climate variability is ongoing—so the connection to evolving climatic risks (and change) needs to be understood.
- The climate science, therefore, needs to change, from scenarios to projections, from snapshots of future time periods to the evolution of the climate system (integrating seasonal forecasting, medium-term El Niño/Southern Oscillation outlooks and greenhouse gas-induced climate change) and from the concentration on means to provision of streams of climate episodes.
- Social processes of adaptation need to be built up from the context of vulnerability (a high number of dimensions), the nature of decision-making (in which climate is only one signal of many) and multilevel, multistakeholder interactions (that embed capacity but also constrain adaptation). Fruitful work on agent-based social simulation is ongoing (see Downing et al., 2001), but many of the essential elements of adaptation are poorly understood and inadequately represented in current models.

8. Response Orientation

Vulnerability assessment grew out of a need to trigger relief, and still keeps a close connection with at least an implied decision-making framework. For example, the US Famine Early Warning System expressly related the (relative) levels of vulnerability to household dynamics and appropriate interventions (Figure 6). Agencies implementing vulnerability assessments seek to distinguish between their own mandate and broader issues, and to collect appropriate information at least for their own decision-making (see the FIVMS website for further information).

Within applications to climate change adaptation, a response orientation requires an actor-oriented approach and recognition as to the relative nature of vulnerability (as noted above). It also requires identification of criteria for evaluation strategies and options. Indeed, multicriteria assessment is similar in construct to vulnerability assessment (selection of key indicators, subjective decision-making, and stakeholder participation). The approach developed by Munda (1995) and the Joint Research Centre (NAIADE, see www.jrc.it) is well worth adopting. Box 3 offers four strategies for agriculture and food security to respond effectively to climate change.

Specific criteria for climate change include the following:

Stakeholders: Should any particular stakeholder or vulnerable group be given priority for targeting adaptive strategies? Are there conflicts over resource use between or among different stakeholders? Would such conflicts affect the ability to design and implement specific adaptive strategies? Long-term adaptation is essentially an intervention in existing resource management. As such, specific projects are unlikely to benefit all stakeholders equally, and some may indeed be worse off—the obvious example is "planned retreat" in coastal zones.

Resilience and effectiveness: Does the adaptation have benefits for a number of objectives and stakeholders? Conversely, does the adaptation only have benefits if climate changes in the expected direction? Is there a critical threshold for adopting the adaptation, or beyond which the response will not be effective? How effective is adaptation in coping with the expected climate change?

Strategic responses: Where the most cost-effective adaptive responses have yet to be identified, fundamental research and stimulation of innovation may be warranted. The role of adaptation in shaping future development needs to be assessed. Costly responses now may limit investment and constrain development.

Level	Conditions	Interventions	Surveillance
Famine	Destitute: coping strategies exhausted.	Emergency relief: food, shelter, medicine, etc.	Monitoring access to relief
Extreme or at risk	Altered production: liquidating productive resources or abandoning preferred means of production in favor of emergency sources of income, employment or production (e.g., migration).	Relief and/or mitigation: support nutrition (e.g., food relief), income (seeds), and assets (fodder)	Intensive community, household, and/or individual assessment through on-site surveys to measure specific needs and resources and identify appropriate responses
High	Depleting assets: liquidating wealth but not yet means of production. Disrupting production: coping strategies have significant individual, household and/or environmental costs.	Mitigation and/or relief: support income and assets (food and cash for work, etc.)	Rapid area and population appraisals: increase information about specific areas and vulnerable groups through site visits and other means
Moderate	Drawing down assets: liquidating less important assets, husbanding resources and minimizing expenditures. Maintaining production strategies: only minor stress-related changes in income strategies.	Mitigation and/or development: support assets (release food stocks, stabilize prices, subsidize fodder, open community grain banks, etc.)	Targeted area and population monitoring: increase information on specific issues and vulnerable groups, but without mobilizing substantial new resources
Slight	Maintaining or accumulating assets: preferred production strategies to cope with seasonal stress and maintain or increase wealth.	Development: long-term strategies to reduce vulnerability	Regular collection of information, primarily for development planning

Figure 6. FEWS vulnerability matrix

Source: Based on a FEWS internal memorandum, 19 April 1991.

Box 3. Adaptation strategies for agriculture and food security

Four strategies offer the kinds of opportunities that could form an effective response to climate risks. The strategies—farm-level agronomy, national economic planning, global agricultural policy and reducing vulnerability to drought—are qualitatively evaluated against the criteria noted in Section 8. Of course, none of the strategies is sufficient on its own—the most vulnerable populations will need a combination of measures, and implementation of even local responses may well require international support.

Common *farm-level adaptive responses* include substitution of agronomic practices, altered inputs and agricultural development. The priority stakeholder is likely to be the smallholder farming sector. Commercial farms would be less likely to need (public) assistance in these sorts of adaptive strategies. On the other hand, these strategies are less likely to be effective for agropastoralists and pastoralism in general. There may be some competition for development assistance, but in general these agronomic packages do not entail serious resource use conflicts between farming groups and others or within farming communities.

All of the adjustments can be implemented relatively quickly, often within a single season. Even investment in soil and water conservation represents a relatively modest investment, compared to the potential risk of climate change. The aggregate evaluation suggests that agronomic improvements are effective, can be readily implemented and have few substantial constraints to their adoption. The short planning horizon suggests that these could be "real-time" responses to (perceived) climate change.

Irrigation schemes may be a special case in considering responses to climate change. They are sensitive to both direct impacts of climate change and to changes in water supply. The design and investment cycle is such that major schemes are expected to be operational for at least several decades, which puts irrigation planning into the time scale of expected climate change.

At the national level, economic policies to promote *agricultural development* focus on maintaining a positive food balance and exports. Three types of strategies can be envisioned. Maintaining strategic reserves allows the government (or marketing bodies) to dampen price fluctuations and release food in emerging crises. Quite large national reserves have been held in the past few decades, and some countries have enough food to meet consumption for a year or more. In the 1990s, these reserves were reduced under structural adjustment agreements. International lending organizations noted that such reserves are costly to maintain and absorb a significant fraction of government

Box 3. Adaptation strategies for agriculture and food security (cont.)

resources. Better monitoring and more timely responses were seen as more efficient ways to cope with food shortages.

An alternative strategy would be to adjust markets and trading conditions to promote private sector responses to climate change and climatic variability. This might take the form of tax incentives for carryover stocks or bonds to smooth income between adverse and good trading years. The stakeholders in such a strategy would be producers, including market traders, millers and agribusiness in general. This kind of strategy would build on current efforts toward reduced trade barriers, with some specific adjustments to accommodate to climate change.

The third realm of national or regional planning is to promote agricultural development in general. This does not require action specifically because of climate change. However, the gap between research and practice is as large as the gap between current yields and agricultural potential in Africa. And, this gap is much less than the implied effect of climate change on yields. The need to adapt to climate change could be used as one argument for fresh initiatives in promoting adaptive agricultural research and development, especially in Africa.

The primary beneficiaries of national economic planning are consumers and commercial producers who depend on markets for food consumption. Market adjustments may entail some tradeoffs between consumers and producers, or between relatively prosperous farmers and vulnerable smallholders who may not have access to inputs and markets. Yet, the potential for multiple benefits is high (except for strategic reserves which are a burden on the economy). Taken as a group, these strategies would be reasonably effective in preparing for climate change.

These strategies can be readily implemented, are not likely to have irreversible impacts (depending on the nature of specific developments) and generally have a strategic role in promoting a resilient economy. Benefits could be realized throughout the economy, although the incremental investment would take 5-10 years to pay off.

Ultimately, prospects for agriculture in vulnerable countries depend on global investment, demand and trade. Indeed, investment in some countries is likely to become more risky with climate change, accelerating the direct impacts of altered weather. At the *global level*, some policies to prepare for climate change may be justified. Suggested strategies range from building strategic reserves, encouraging free trade and transferring agricultural technology.

Box 3. Adaptation strategies for agriculture and food security (cont.)

The arguments for building global strategic reserves, both for major foods and of financing, follow the same argument as that at the national level. Climate change may require additional trade to smooth out fluctuations in national production. Maintaining international prices within acceptable limits would benefit poorer countries who might not be able to afford large imports in times of scarcity (as might occur in a replay of the 1980s drought on the 1972 international scarcity of grains). It should be more effective to hold reserves at the global level, or shared among regional trading partners, rather than each country seeking to buffer its internal production. In the transition toward a new climate, such an international capacity to prevent food deficits becoming survival emergencies appeals to humanitarian goals of ending famine and reducing hunger.

Encouraging free trade between countries should stimulate agricultural markets in regions with a comparative advantage. This may be a major benefit to some countries and a significant cost to others, as the impacts of climate change alter traditional markets. In principle, free trade allows national surpluses and deficits to be accommodated more efficiently. Supply and price fluctuations are thus buffered at the global level, widening the potential pool of responses to climate change. Free trade, of course, is already on the international agenda and little further encouragement is required. However, some incentives to the private sector to absorb additional risks may be required.

The most costly and long-term strategy proposed in this review is to develop international mechanisms to promote agrotechnology transfers to developing countries. An initial agreement might focus on basic foodstuffs: wheat, rice and maize. International agencies might license new technologies developed by biotechnology firms for dissemination and use in developing countries. Adaptation and abatement might be explicitly linked, with a requirement that beneficiaries have agreed to limit greenhouse gas emissions. Funding requirements would be significant, although connections to emission taxes could be made.

The immediate beneficiaries of global linkages are commodity brokers and private companies, although aid and government agencies have strong interests. As for national policies, some resource allocation issues may imply tradeoffs between regions, commodities and farming populations. Global policies should be highly effective, with a relatively low specificity to climate change or dependence on specific climate scenarios.

Box 3. Adaptation strategies for agriculture and food security (cont.)

The planning horizon is generally five years, with benefits realized somewhat later. Except for agrotechnology transfer, there are few constraints to implementing global policies (other than bureaucratic inertia and funding).

Drought hazard and vulnerability are current risks and likely to be the most damaging locus of impacts of climate change. Concerted action is required in three areas: mitigation to reduce vulnerability, monitoring of drought and vulnerability, and preparedness to respond effectively to emerging crises. Considerable progress has been made in the past decade, and a further decade of development might reap substantial rewards in efforts to eliminate widespread famine and enhance livelihood security, at least in times of drought.

The priority stakeholders should be the most vulnerable socioeconomic groups affected by drought crises, although many levels of local, national and international actors are required to implement drought monitoring, mitigation and emergency responses.

Drought policies provide multiple benefits to the extent that they contribute to development in general by reducing investment risk. However, crisis interventions can be counterproductive if they create a dependency syndrome or compete for resources from other activities with longer term benefits.

Reducing risk can have strategic importance. Increased monitoring and response capabilities should improve the ability to respond to long-term climate change and to economic management of the agricultural sector in general. With foresight, crises can be used to promote sound resource policies, although this remains the anomaly in most of the world rather than the norm.

Drought monitoring, mitigation and preparedness take time to develop and implement. Once operational the benefits are immediate, although they are significant only during times of potential crisis. There are few irreversible impacts and the initial investment can be fairly modest.

Constraints include the need for sustained information collection, processing and reporting, often requiring significant development of technologies and organizations. While most planners and vulnerable populations agree that drought hazard planning and reduction are desirable, the lack of reliable systems implies further social, economic and political constraints. One that is commonly cited is the short attention cycle—drought planning peaks about a year after the drought and is forgotten until the next crisis. If drought becomes more frequent, this constraint may be overcome.

Timing: Is there a gap between the implementation of adaptation strategies and the realization of their benefits? What is the planning horizon required to design and implement the adaptation? For how long is the adaptation useful? Matching the timing of an adaptive response and its benefits may need to consider the possibility of irreversible impacts and option values.

Economic evaluation: Contentious issues of discount rates, environmental quality valuation and equity need to be addressed in balancing costs and benefits. The initial investment may be a constraint, either because of poverty and lack of credit among some stakeholders or because the expected return is low compared to other economic investments. The difference between public and private objectives and economic decision-making may be important in such cases. While most guidelines recommend "low cost" adaptations, the definition of low cost depends on each stakeholder and their opportunity costs. The opportunity cost in developing countries may be quite high—arguing for a relatively high discount rate. The return on investment depends on the timing of the benefits. The benefits may not accrue to the stakeholders that make the investment. The beneficiaries are future generations, or other current social and economic groups—common issues of inter- and intragenerational equity.

Constraints: Some strategies will be effective only if they are widely adopted, with relatively efficient means of dissemination and maintenance. Is information about the strategy, its utility and means of implementing it lacking? Is the adaptation technically reliable, or does it require a level of technical development that is not available for all stakeholders? Do institutional, behavioral, cultural or political conditions influence the range of choice and adoption?

9. Standards

Recently, a consortium has worked to create and implement a series of standards for disaster relief (Sphere Project, 2000). While not universally accepted, the process and experience around the world is proving helpful to discuss ways of promoting effective disaster aid. At the same time, methods in famine early warning systems have converged, or at least reflect over a decade of review and comparison (see the FIVMS web site at www.fivms.org). Some of the methods have been generalized (for example, the poverty mapping initiative, www.povertymap.net). These efforts have been successful in part due to a dominant paradigm for the past decade—expressed in Sen's (1981) entitlements theory and the more general approach of livelihood security.

The experience in climate change impacts research is less coherent. The early dominance of the "seven steps" (Carter et al., 1994) tended to narrow the range of research, and often led to inappropriate methods that ignored present vulnerability. Certainly, the progression from impacts to adaptation has challenged the IPCC approach. (Indeed, the innovation in adaptation science and practice has largely come from outside the IPCC formal structures.) Intercomparison of some models has been undertaken by the Global Change and Terrestrial Ecosystems programme of the International Geosphere-Biosphere Programme. But this has been limited to first-order models, notably crop-climate and vegetation models. A systematic comparison of higher order models (e.g., sectoral assessments and vulnerability mapping) is still lacking. Even a documented data set of climate change impacts does not exist—anecdotal experience suggests that the choice of impact model is at least as important in defining the final results as are the climate and socioeconomic scenarios.

And the climate change community does not appear to be very close to adopting standards, or even a common paradigm and analytical language.

So, what would be essential in a vulnerability/adaptation science? The approach is related to the grander notion of a new sustainability science (NRC, 1999), while more clearly targeting leading situations of vulnerability that are likely to be affected by global change and should be highest priority for adaptation—for present humanitarian and development concerns as well as reducing the risks of climate change. The main characteristics are as follows:

Integration: Vulnerable socioeconomic groups integrate processes (for example, changes in natural resources and economic structures) and shocks (the episodic threats, for instance, of drought and recession). What are they exposed to? On what time scales? Where?

Multilevel: The context of the vulnerable is essential, beyond households and communities. Adaptation requires matching effective action across relevant scales.

Structure and agency: The actions of the vulnerable must be placed within the resource, social and economic structures that allocate resources, regulate exchanges and empower access.

Complexity and constraints: Processes of vulnerability are diverse, interact in unanticipated ways and rarely are informed by adequate data.

Adaptation and evolution: Pathways from vulnerability to capacity, or the evolution of new vulnerable groups require a process-oriented science.

Participatory: The development mantra of participation has rarely been translated into effective scientific paradigms, although some experiments with famine early warning systems are informative.

Linking development, disasters and long-run climate change: Coping with existing climatic risks is a first step, and provides insight into mechanisms for adapting to climate change.

Targeting and monitoring: Methods to decide priorities, to differentiate between strategies and to monitor success need to be developed, and adopted.

While there are still significant differences of opinion regarding the best paradigm, the toolkit for climate change impacts studies has been expanded (for example, see Willows et al., forthcoming). However, matching the potential methods to the goals and building blocks in an assessment remains poorly documented. Recent discussions suggest that two of the key characteristics of defining the problem are climate sensitivity and level of decision-making:

- The relative dominance of climate sensitivity in the system being investigated. For example, the issue at hand may resolve around the supply of irrigation water in a protected catchment with few other users or uses. In this case, the variability in water supply is likely to be strongly linked to climatic variability. On the other hand, if the key issue is household vulnerability to drought (in the same irrigation scheme), other, largely nonclimatic, factors will need to be investigated, possibly reducing the relative importance of drought to a triggering mechanism.
- The level of decision-making or social organization taken as the unit of exposure or analysis. This may range from a single actor (whether an individual, organization or omniscient decision maker) to multistakeholder negotiations over responses to climatic variations (and other social and economic perturbations). The scale of decision-making generally broadly matches the level of decision-making—from operational to strategic policy, although these may be distinct characteristics of the issue.

Figure 7 depicts these two characteristics and suggests the kinds of methods that are appropriate for different types of problems. Of course, more than one method is generally desirable, and the methods may combine to give specific insight. The boundaries between the relevant domains of each method are quite fuzzy—and it is difficult to discriminate a priori between the most appropriate methods for a specific study.

10. Conclusion

Adaptive measures to promote food security in the face of climate change are known, and methods for reviewing potentially effective strategies linking across scales are well advanced, even if not well implemented. Three cautions, however, are remarkable:

- First, the implication of displacement in agro-ecological potential is an increase in world food trade. Under some scenarios of emission pricing, this may be costly and poor countries and consumers may not be able to command food at world prices that include high transport costs.
- Second, within regions and nations, the engine of food security for rural households is investment in agriculture, and this may be problematic in a world where more crops can be grown in higher latitudes or where climate forecasts suggest increased risks (as in semi-arid tropical regions).
- Third, livelihood security is a "good thing" but even current modest international development targets are unlikely to be achieved in many places. Linking global change and poverty alleviation makes sense, but neither comes with a guarantee to redress fundamental disparities in world economic and political systems.
- Ten years ago the Oxford Conference on Climate Change and World Food Security (Downing, 1996) posed the question: How significant is the risk of climate change for regional collapse? The workshop concluded that climate change over the next decades was unlikely to seriously affect global food production. We also believed that individual livelihood security would come under increasing stress in some locales, although those might not be easily identified. We were not confident in our ability to analyze the risks at the intermediate

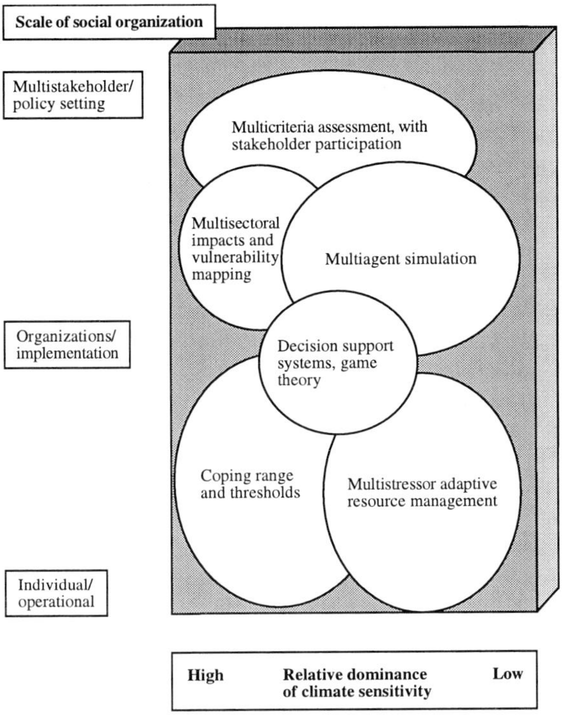

Figure 7. Climate change vulnerability and adaptation methodologies

scale of regions. And, we are still not able to answer this fundamental challenge to our science.

References

Carter, T. R., Parry, M. L., Nishioka, S. and Harasawa, H. (1994), *IPCC Technical Guidelines for Assessing Climate Change Impacts and Adaptations.* Environmental Change Unit, Oxford, United Kingdom.

Downing, T. E., ed. (1996), *Climate Change and World Food Security.* Springer, Berlin.

Downing, T. E. (2002), Protecting the vulnerable: climate change and food security. In *Managing the Earth,* Briden, J. C. and Downing, T. E., eds. Oxford University Press, Oxford, UK, 157-178.

Downing, T. E. (forthcoming), Linking sustainable livelihoods and global climate change in vulnerable food systems. *Die Erde.*

Downing, T. E., Moss, S. and Pahl-Wostl, C. (2001), Understanding climate policy using participatory agent-based social simulation. In *Multi-agent Based*

Simulation: Second International Workshop (MABS 2000). Lecture Notes in Computer Science, Moss S. and Davidson, P., eds. Vol. 1979, 198-213.

Downing, T. E., Eyre, N., Greener, R. A. and Blackwell, D. (1996), *Full Fuel Cycle Study: Evaluation of the Global Warming Externality for Fossil Fuel Cycles with and without CO_2 Abatement and for Two Reference Scenarios.* Report to the International Energy Agency, Greenhouse Gas R&D Programme. IEA, Cheltenham, UK.

Downing, T. E., Ringius, L., Hulme, M. and Waughray, D. (1997), Adapting to climate change in Africa: prospects and guidelines. *Mitigation and Adaptation Strategies for Global Change* **2**, 19-44.

Downing, T. E., Butterfield, R., Cohen, S., Huq, S., Moss, R., Rahman, A., Sokona, Y. and Stephen, L. (2001), *Vulnerability Indices: Climate Change Impacts and Adaptation.* UN Environment Programme, Nairobi.

Fischer, G., Shah, M., van Velthuizen, H. and Nachtergaele, F. O. (2001), *Global Agro-ecological Assessment for Agriculture in the 21st Century.* IIASA, Laxenburg and FAO, Rome.

Fischer, G., Shah, M., van Velthuizen, H. and Nachtergaele, F. O. (2002), *Global Agro-ecological Assessment for Agriculture in the 21st Century: Methodology and Results.* IIASA, Laxenburg and FAO, Rome.

Jones, R. N. and Page, C. M. (2001), Assessing the risk of climate change on the water resources of the Macquarie River catchment. In *Integrating Models for Natural Resources Management across Disciplines, Issues and Scales* (Volume 2), Modsim 2001 International Congress on Modelling and Simulation, Ghassemi, F., Whetton, P., Little, R. and Littleboy, M., eds. Modelling and Simulation Society of Australia and New Zealand, Canberra.

Munda, G. (1995), *Multicriteria Evaluation in a Fuzzy Environment.* Contributions to Economics Series. Physica-Verlag, Heidelberg, Germany.

NRC (1999), *Global Environmental Change: Research Pathways for the Next Decade.* National Research Council. National Academy Press, Washington, DC.

Sen, A. K. (1981), *Poverty and Famines.* Clarendon Press, Oxford, UK.

Sphere Project (2000), *Minimum Standards for Disaster Response.* Oxfam, Oxford, UK.

Sphere Project (2001), Report of an Inter-Agency Workshop to Discuss Minimum Standards for Food Security in Disaster Response, Oxford, 2-3 July. Oxfam, Oxford, UK.

Stephen, L. and Downing, T. E. (2001), Getting the scale right: a comparison of analytical methods for vulnerability assessment and household level targeting. *Disasters* **25**, 113-135.

UNDP (1999), *Human Development Index.* United Nations Development Programme, New York.

Willows, R. I., Reynard, N., Connell, R. and Meadowcroft, I. (forthcoming), *Climate Adaptation: Risk, Uncertainty and Decision-Making* UKCIP Technical Report. U.K. Climate Impacts Programme, Oxford, UK.

6. Assessing Vulnerability in the Context of Changing Socioeconomic Conditions: A Study of Egypt

Gary Yohe
Department of Economics
Wesleyan University, Middletown, Connecticut, USA

Kenneth Strzepek
Civil, Environmental and Architectural Engineering
University of Colorado, Boulder, Colorado, USA

Tammy Pau and Courtney Yohe
Department of Economics
Wesleyan University, Middletown, Connecticut, USA

A range of "not implausible" climate scenarios is superimposed on a range of similarly "not implausible" socioeconomic scenarios for Egypt to explore the role of adaptation in reducing vulnerability and to illustrate an approach derived from the determinants of adaptive capacity. The numerical results depend on context and model specification, but several robust qualitative insights are supported. Adaptation can make a significant difference on a macro scale, especially for pessimistic climate scenarios. Socioeconomic context matters in determining adaptive capacity, and inefficient investment can diminish that capacity. The value of information that supports early differentiation between two strikingly different climate futures can be significant. Moreover, early preparation can be important because macro-scale adaptation can involve capital reallocation between sectors in anticipation of large future investments in adaptation infrastructure. Planning for bad news and adapting to good can be a better choice than the other way around, but working to expand the potential of some options to increase adaptive capacity can create rigidities or cause systems to under-prepare for other options. These omissions can reduce the capacity to cope with more extreme climate futures because the first set of adaptations may be overwhelmed even as more efficacious adaptations become less feasible.

1. Introduction

Strzepek et al. (2001) described a process by which nine "not implausible" climate scenarios were selected for Egypt. Their selection was the first step of a project designed ultimately to conduct detailed integrated assessments of the impacts of a range of climate changes expressed as scenarios across a range of similarly "not implausible" socioeconomic scenarios. Here we report on progress in defining aggregate portraits of the requisite diverse set of socioeconomic scenarios and in exercising those scenarios to investigate the economic implications of micro- and macro-scale adaptations. Impacts on agricultural and nonagricultural production of climate-induced reductions in flow along the Nile are the focus of our attention, but this is not an Egyptian case study. It is, instead, a methodological piece that uses an Egyptian context to explore how the fundamentals of economic analysis might be applied to adaptation issues. It turns out, however, that the Egyptian context is particularly well suited for this task.

A rich diversity of possible socioeconomic scenarios was created by spanning a range of not-implausible representatives of a few macro-scale determinants of adaptive capacity. Section 2 sets the stage by reviewing those determinants within a pedagogical model that sees vulnerability to climate change and climate variability as a function of exposure, sensitivity, and adaptive capacity. Section 3 then reviews the representative climate scenarios from the earlier work by Strzepek et al. (2001) that display a range of Egypt's exposure to climate change. Indeed, they include one scenario in which flow in the Nile actually increases, but eight less optimistic alternatives range from modest reductions to declines that would appear to be quite severe. Section 4 follows with a description of a Ramsey-style aggregate growth model that was designed specifically to accommodate investigations of the relative efficacy of municipal and industrial recycling, drip irrigation and groundwater pumping in relieving climate induced stress on macroeconomic activity and food self-sufficiency. Details of the model are provided in an appendix to this paper. As reported to us by the Minister of Water and Irrigation, macroeconomic vitality and food self-sufficiency are both explicit policy objectives of the Egyptian government; and these three adaptations are under active consideration in support of both.

A collection of representative socioeconomic scenarios that extends across a wide range of "not-implausibility" defined by variation the macro-scale determinants of adaptive capacity for these options is described in the Section 5. It is within these scenarios that Egypt's sensitivity to the climate scenarios described in Section 3 are explored in Section 6 with careful attention being paid to the potential efficacy of the three possible adaptations. Section 7 then displays the utility of the conceptual methodology described in Section 2 by offering some interpretative results drawn from that structure. Section 8 responds to a hypothesis that was derived from the adaptation results of Section 7 by exploring the value of information in implementing adaptation before a closing section offers some contextual conclusions. Notwithstanding the specific context from which they were drawn, we expect that these conclusions hold considerable validity beyond the boundaries of the Egyptian illustration.

2. Adaptive Capacity

The Intergovernmental Panel on Climate Change (IPCC) envisioned a broad relationship between vulnerability, sensitivity, and adaptive capacity in its Third Assessment Report (McCarthy et al., 2001). As they reviewed adaptation and adaptive capacity in this context, the authors of Chapter 18, on "Adaptation in the Context of Equity and Sustainable Development", as well as the authors of other sector and regional chapters, came to recognize that this relationship was complex, location specific and path dependent. Indeed, many would now contend that any subsequent analysis that did not recognize regional diversity in development trajectories, uncertainty in climate futures and the potential for adaptation would be suspect.

To be more specific, the authors of Chapter 18 concluded that adaptive capacity varies significantly from system to system, sector to sector and region to region. Indeed, they noted that the determinants of adaptive capacity include a variety of system, sector and location-specific characteristics:

- the range of available technological options for adaptation
- the availability of resources and their distribution across the population
- the structure of critical institutions, the derivative allocation of decision-making authority and the decision criteria that would be employed (i.e., governance)
- the stock of human capital including education and personal security
- the stock of social capital including the definition of property rights
- access to risk spreading mechanisms

- the ability of decision makers to manage information, the processes by which these decision makers determine which information is credible and the credibility of the decision makers, themselves
- the public's perceived attribution of the source of stress and the significance of exposure to its local manifestations.

Many of these determinants cannot easily be quantified, but working through their content from the bottom up or from the top down can nonetheless uncover practical insights that can inform our understanding of how adaptation might diminish vulnerability.

The Egypt work reported here takes a top-down approach based on the observation that many of the determinants of adaptive capacity operate on macro scales in which national or regional factors play the most significant role. While the set of available, applicable, and appropriate technological options (Determinant 1) for a given exposure at a particular location might be defined on a micro scale, for example, the complete set of possible remedies for a national response should have macro roots. Determinants 2 through 6 should all have large macro components even though their micro-scale manifestations could vary from location to location or even from adaptation option to adaptation option. Resources (Determinant 2) could be distributed differently across specific locations, but adaptive capacity may be more sensitive to larger scale issues that determine the availability of resources across an entire nation. The essential questions here focus on whether sufficient funds are available to pay for adaptation and whether the people who control those funds are prepared to spend them on adaptation. Macro-scale and even international institutions (Determinant 3) could also certainly play a role in determining how decisions among various adaptation options might be made and who has access to the decision-making process. Adaptation projects will be directed towards improving well-being measured against domestically determined objectives. The stocks of human capital and social capital (Determinants 4 and 5) could be locally idiosyncratic, as well, but their local manifestations would most likely be driven in large measure by macro-scale forces such as national education programs and the efficiency of public investment. Access to risk spreading mechanisms (Determinant 6) usually evokes notions of insurance and monetary compensation after the fact; by their very nature, these are macro in scale. There are, however, many instances in which adaptation before the fact can function as a physical "insurance polity" on a micro scale—a process by which exposure and/or sensitivity might be diminished in the face of an uncertain future. Finally, Determinants 7 (informational management) and 8 (attribution of signals of change) for national adaptations must have macro-scale foundations even if their force is derived from local vulnerabilities.

3. Climate Scenarios for Egypt

Panels A and B of Figure 1 display nine representative climate scenarios in terms of flow into Lake Nasser and the area of upstream swamps in the Sudan, respectively; they represent the primal exposure of Egypt to climate change. Each was driven by specific assumptions about greenhouse gas and sulfate emissions, climate and sulfate aerosol sensitivities and the results of some specific global circulation model, but each was selected for its representative value. Taken together, these nine scenarios span a range of outputs produced by running COSMIC for rainfall and temperature for nine upstream countries through a hydrological model authored by Yates and Strzepek (1998). Decadal markers between 2000 and 2100 are depicted, and swamp area (Panel B) is included because draining that swamp could have been a macro-scale adaptation along water-scarce futures. Given the political friction that would be created by draining a Sudanese resource to sustain Egyptian economic activity, it is perhaps good news that the swamp would, under these scenarios, not be available.

4. An Economic Model for Egypt

A modification of the classical Ramsey analysis of optimal economic growth under certainty provided the modeling context for describing how Egypt might move into the future. It is described in some detail in the appendix to this paper. Blanchard and Fischer (1989) and Barro and Sala-i-Martin (1995) offer general discussions of the fundamental Ramsey construction, but a nonlinear formulation of the classical Ramsey model developed by Lau et al. (2002) was employed. The primal formulation was based on an explicit representation of utility for a representative household that depended on per capita consumption. The social planner maximized its present value subject to the constraint that output in period t was either consumed or invested. It is convenient to think of the production function exhibiting constant returns to scale in capital and a second factor whose supply would be exogenously

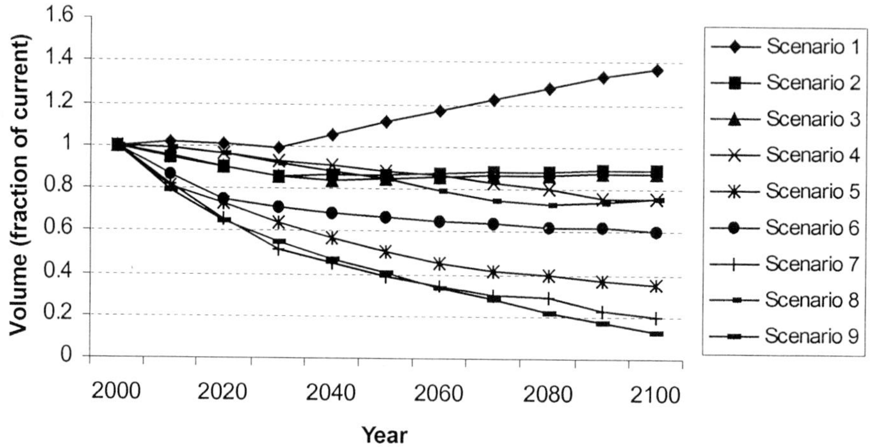

Figure 1; Panel A. Flow into Lake Nassar (current flow = 1)

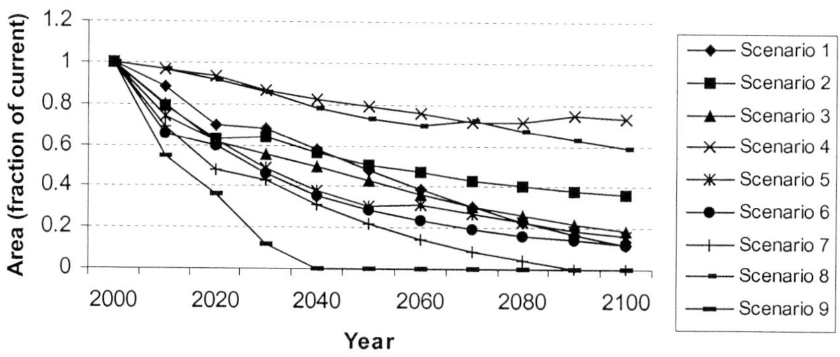

Figure 1; Panel B. Area of upstream swampland (current area = 1)

specified. The capital stock in each year equaled the capital stock at the start of the previous year less depreciation plus investment in the previous period.

Several major modifications were made to this model to represent the Egyptian economy more accurately. Production was, first of all, divided into two major sectors with sector-specific capital. The agricultural sector produced only food for domestic consumption. The nonagricultural sector produced an output that could be invested to produce capital, consumed or exported to pay for imported food. Second, the single household differentiated two types of consumption goods: food and a nonfood consumable. The underlying accounting process noted two types of food. The first sustained a minimum caloric requirement of 2100 cal/per capita/day; it was included as a constraint in the model. The second built supplemental caloric intake up to 1100 cal/per capita/day. Supplemental calories were preferable, on a declining scale, to nonfood consumables. This complication added a food balance constraint, and changed the modified objective function.

The classic Ramsey model does not allow for trade, of course, but Egypt is currently only 70 percent food self-sufficient. Assuming a closed economy would therefore be inconsistent with current and, in all likelihood, future realities. The model was modified to allow nonagricultural output to be exported for equal the amount of food imports subject to the terms of trade for Egyptian nonagricultural output on the world market. The terms of trade parameter was exogenously specified so that a balance equation for production and consumption could be specified.

Water is, of course, an essential factor of production in the fully irrigated Egyptian agricultural sector as well as an important factor in hydroelectricity production, some industries and transportation. Water was therefore included in the model in a dynamic Leontief fashion. Two different but constant rates of technological progress were specified for each of the two sectors so that water use per unit output in either sector would decline over time. Water was also demanded for domestic consumption, and this use received the highest priority. Domestic water use was modeled as a function of income expressed by GDP/capita. Domestic water use typically increases at a very steep rate until income reaches approximate $2,000 per capita (constant 2000 US dollars) then it becomes almost constant. Taking these three demands together, a water balance equation was imposed on the model.

Finally, it should be noted that specifying exogenous target levels of food self-sufficiency frequently resulted in infeasible solutions when the potential impacts of climate change on Nile flows and agricultural water requirements were extreme. To avoid this problem, the level of food self-sufficiency was modeled as an endogenous variable to be determined by the model. Private communications with government officials and Egyptian academics reveal that maximizing food self-sufficiency is policy goal of the Egyptian government. To model such a policy, the objective function was modified to reflect disutility derived from importing food equal to consumption of nonagricultural good. This structure placed additional weight on minimizing food imports because each unit of food imports caused a double loss in utility.

5. Selecting the Socioeconomic Scenarios

The selection process for socioeconomic scenarios for detailed analyses of adaptations began by running the Ramsey model for more than 600 combinations of nine climate scenarios, two alternative population futures, and two alternative specifications of four underlying socioeconomic parameters. Table 1 summarizes the variables that determined the range of possible socioeconomic context. Two population scenarios allowed population to stabilize by the middle of the 22nd century at 1.5 to 2.5 times current levels. The resulting range spanned projections offered by various sources using different assumptions about family size, longevity, and cultural perspective.

The determinants of adaptive capacity described in Section 1 highlight the potential significance of resource availability (and distribution) as well as the ability of decision makers to allocate those resources effectively. Strzepek et al. (2001) reinforced this message by noting different capacities to adapt under high and low capital futures. Table 1 shows that variation across high or low paths for three critical parameters reflected the import of these insights with considerable richness. Nonagricultural productivity growth, for example, assumed high or low values of 2 percent or 1 percent per year. Agricultural yields were similarly given high or low trajectories with rates of 1.5 percent or 0.5 percent per year. Finally, the efficiency of investment was assumed to be high (normal) or low by assuming values of 1.0 or 0.8, respectively. The low value in this case meant that one unit of output devoted to investment would, by virtue of misallocation by government planners and/or associated second-best allocations by private

Table 1. Parameterization of "not-implausible" climate and socioeconomic futures

Parameter	Description
Climate	Nine distinct scenarios depicted in Figure 1
Population	Scenario 1: Growth to 1.5 times current levels by 2100
	Scenario 2: Growth to 2.5 times current levels by 2100
Technological change in the nonagricultural sector	High: 2.0% per year; Low: 1.0% per year
Technological change in the agricultural sector	High: 1.5% per year; Low: 0.5% per year
Terms of trade	High: 1 for 1; Low: 0.8 for 1
Investment efficiency	High: 1 for 1; Low: 0.8 for 1

investors, increase the "effective" capital stock by only 0.8 units. Finally, Egypt's future trading position in the world market will be a critical determinant of the availability of resources. The terms of trade were therefore assumed to be favorable or unfavorable by assuming high or low values of 1.0 or 0.8, respectively. Low terms of trade meant, for example, that 1.0 unit of consumable goods traded on the world market would produce 0.8 units of imported food.

Figure 2 displays the results of these runs in terms of an index of Egyptian food self-sufficiency and total food plus consumable good consumption in 2067 drawn from more than 600 different combinations of variables identified in Table 1. We chose 2067 because it reflects a point in the relatively distant future by which time the nine climate scenarios had, for the most part, diverged. The implications of climate and socioeconomic circumstances were therefore fully represented. The food self-sufficiency coefficient reflects the proportion of total food consumption supported by domestic food production. It was chosen as an important indicator of Egypt's future because of the importance placed on food security by the Egyptian government; this policy objective was identified in Strzepek et al. (2001) as a critical differentiating characteristic of future economic vitality. The sum of food and consumable consumption was chosen, as well, because it reflects

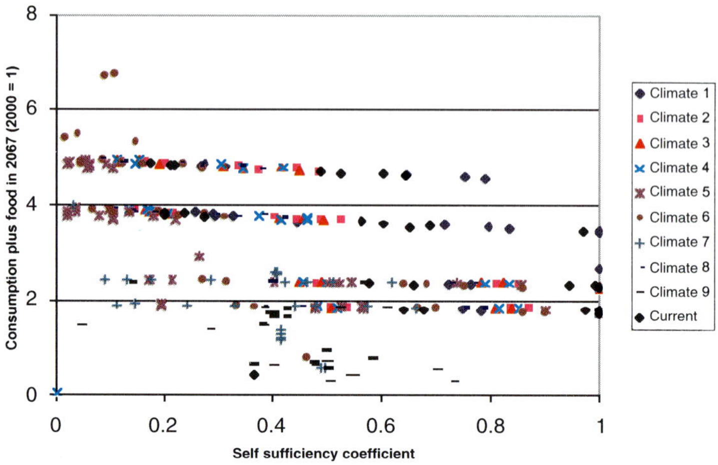

Figure 2. Distribution of not-implausible futures displayed in terms of an index of consumption goods and food in 2067 (2000 level = 1) and a self sufficiency coefficient (the proportion of food consumption grown domestically)

critical components of the determinants of domestic welfare; it is reflected as a multiple of the level achieved in 2000. Finally, there is nothing special about range of results produced for 2067. Other years, from roughly 2030 through 2100, produced ranges that were entirely comparable to those depicted in Figure 2.

Linear patterns are clear in Figure 2, and investigating their sources made it relatively easy to select a manageable number of representative scenarios. The first step sorted the results by climate regime, and six groupings emerged. The patterns for climate scenarios 1, 5, and 6 seemed to be unique, for example, but points indicating various socioeconomic futures for scenarios 2 and 3 tended to bunch together. Climate scenario 3 was selected to reflect these possibilities. Scenarios 4 and 8 also displayed similar patterns, so scenario 8 was selected. Finally, scenarios 7 and 9 displayed results with little diversity; and scenario 9 was selected to carry those futures forward. Placing the runs from these climate scenarios into six groups also produced patterns—different clusters for different population scenarios. Each showed varying ranges of food self-sufficiency and total consumption for across the various socioeconomic specifications. Some of these ranges were large; others were quite small. The limits of each, though, could be captured by the same combination of parameterizations; they are identified in Table 2 as (socioeconomic) scenarios A through F. Figure 3 places these 36 representations into the context of Figure 2 to show the considerable degree to which, taken together, they span the diversity of the original larger set of possible futures.

The underlying details for six different socioeconomic scenarios in Table 2 reveal our choice of two to represent diversity across the lower population scenario; four others accomplished the same task for the higher population future. Notice, in the latter case, that scenarios C and D both show high efficiency in investment, high productivity growth in the nonagricultural sector and unfavorable terms of trade; they are differentiated solely by different assumptions about productivity growth in the agricultural sector. Scenarios E and F, meanwhile, with low investment efficiency and favorable terms of trade, are differentiated by pair-wise deviations in assumptions about productivity growth in the two sectors. The same socioeconomic scenarios were chosen for each climate scenarios to make comparisons easier to interpret; but Figure 3 shows that they nonetheless captured the diversity of "not-implausible" futures reflected by the more than 600 combinations displayed in Figure 2. Figure 4 offers a different portrait of this diversity by tracking time series of GDP estimates for the higher population assumption along three climate scenarios—one with modest climate change (scenario 3), one with more severe change (scenario 6) and one with extreme change (scenario 9).

Table 2. Details of the representative socioeconomic scenarios

	Population	Nonagricultural productivity	Agricultural productivity	Investment efficiency	Terms of trade
Scenario A	Low	High	Low	High	High
Scenario B	Low	High	Low	Low	Low
Scenario C	Higher	High	High	High	High
Scenario D	Higher	High	Low	High	High
Scenario E	Higher	High	Low	Low	Low
Scenario F	Higher	Low	High	Low	Low

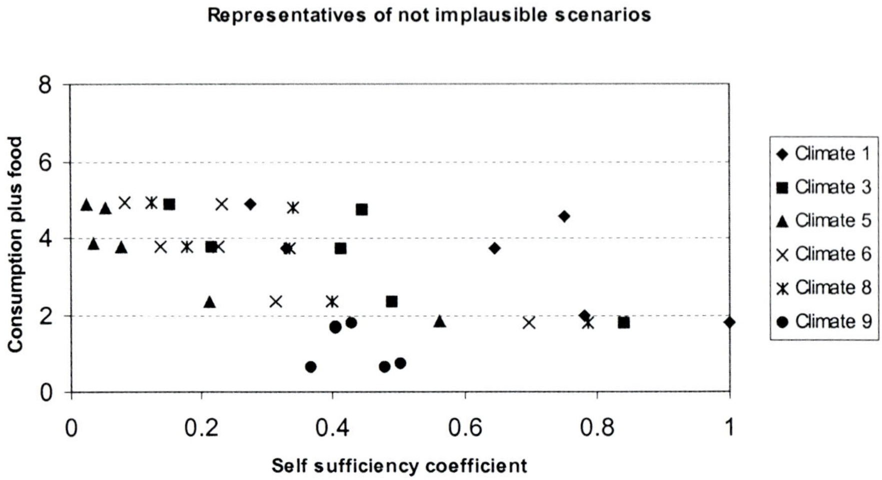

Figure 3. Distribution of not-implausible futures for representative scenarios displayed in terms of an index of consumption goods and food in 2067 (2000 level = 1) and a food self sufficiency coefficient (the proportion of food consumption grown domestically)

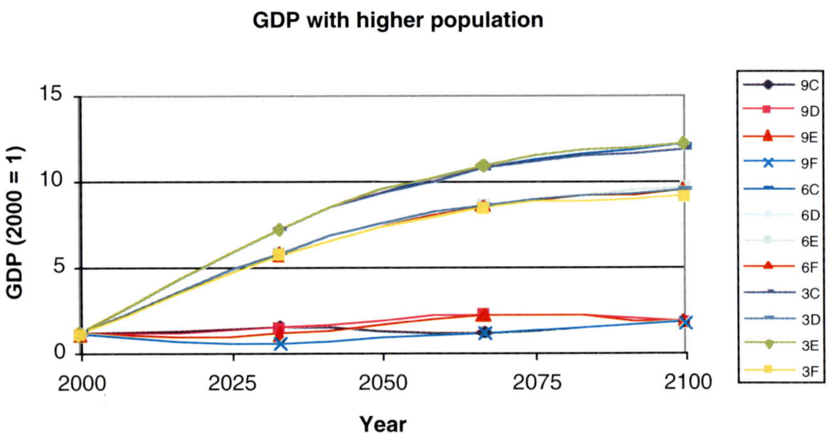

Figure 4. GDP trajectories for Egypt along representative climate and socioeconomic scenarios with the higher population growth

6. Adaptation across the Climate cum Socioeconomic Scenarios

A careful review of Figure 1 sets the context for thinking about the ramifications of alternative socioeconomic scenarios, especially with a view towards framing experiments designed to investigate the role of possible macro-scale adaptations. Three alternatives have been incorporated into the model. It was assumed that two alternatives, municipal recycling and drip irrigation, could be adopted on a micro scale with relatively small investments in equipment and infrastructure. In fact, these modeled to be autonomous adaptations that were adopted as needed along any climate scenario. It was also assumed that a third adaptation, groundwater pumping from an enormous aquifer located under the Sahara Desert, could only be adopted on a macro scale with massive public investment from the federal government. A project of this sort is under active consideration by the federal government. It was, therefore, not ruled out of hand for a government that is prone to underwriting massive public projects.

Scenario 1 could produce favorable outcomes from climate change as long as potential floodwaters could be diverted into vacant and domestic regions of the Sahara Desert. Flow into Lake Nasser would be stable along this scenario through 2030 and then climb over the next 70 years. Since flow would be 40 percent higher by 2100, though, some attention should be paid to the capacity of the Aswan Dam to accommodate this scenario sometime in the next half century.

Scenarios 2 and 3 would be relatively benign. Flow would fall by roughly 8 percent by 2030, but that level would be maintained across the rest of the 22nd century. Scenarios 4 and 8 would also portend modest climate change with a gradual decline by 2100 of approximately 12 percent. Scenario 6 offers the first portrait of serious shortfall in Nile flow. Flow would fall by 25 percent by 2025, thereby tracking even the worst climate outcomes over the near-term; but it would decline only gradually thereafter for a total reduction of 40 percent by 2100. Scenario 5 tracks scenario 6 through 2025, but subsequent reductions would be more severe. Indeed, flow into Lake Nasser would be 55 percent and 65 percent lower then the present value by 2067 and 2100, respectively. Finally, scenarios 7 and 9 would produce the worst outcomes in terms of climate change. Near-term reductions of 30 percent by 2025 are not much worse than 5 and 6; but flow falls by 75 percent by 2067 and by 80 percent around the turn of the next century.

The small points highlighted on the various panels of Figure 5 indicate the outcomes for representative socioeconomic scenarios with the higher population assumption for four representative not-implausible climate futures absent any adaptation; they replicate the corresponding points depicted in Figure 3. Panel A, for example, displays outcomes in 2067 with no climate change. Panel B therefore indicates that the scenario 3 could support similarly high levels of consumption and food intake, but the potential for food self-sufficiency would be diminished somewhat. Panel C depicts scenario 6. It again shows that comparable levels for consumption and food intake could be sustained even with serious shortfalls in flow into the lower Nile, but Egypt would be afforded even less flexibility in a restricted ability to achieve increased food self-sufficiency. Panel D, though, shows that there are limits to the ability of the Egyptian economy to cope routinely with flow reductions. The summary representations show that the worst climate possibilities would severely limit consumption and food intake and would dramatically restrict Egypt's ability to manipulate its food import level.

The large points drawn on the last three panels of Figure 5 portray outcomes given adaptation under two assumptions dictated by the perfect foresight of the Ramsey modeling approach:

- The rational, forward looking household sector would optimally install the infrastructure as required to phase in municipal recycling.
- The government would undertake the extraordinary capital investment required to deliver pumped groundwater from beneath the Sahara Desert.

Panel B, for example, links outcomes in 2067 with and without adaptation for four socioeconomic scenarios for the middle population trajectory along climate scenario 3. Outcomes without and with adaptation for each socioeconomic scenario are connected with dotted lines; and the adaptation outcome is identified by the larger endpoint. Notice that the value of adaptation is seen most clearly in terms of increased food self-sufficiency, sometimes at the expense of some economic activity. Perhaps more surprisingly, a comparison with Panel A suggests that the impacts of the modest climate change of scenario 3 can be accommodated entirely by adaptation.

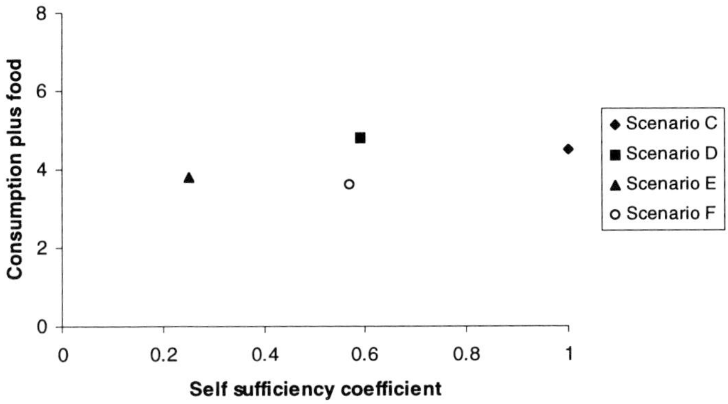

Figure 5; Panel A. Outcomes in 2067 along representative socioeconomic scenarios with the higher population trajectory displayed in terms of an index of consumption goods and food in 2067 (2000 level = 1) and a food self sufficiency coefficient (the proportion of food consumption grown domestically).

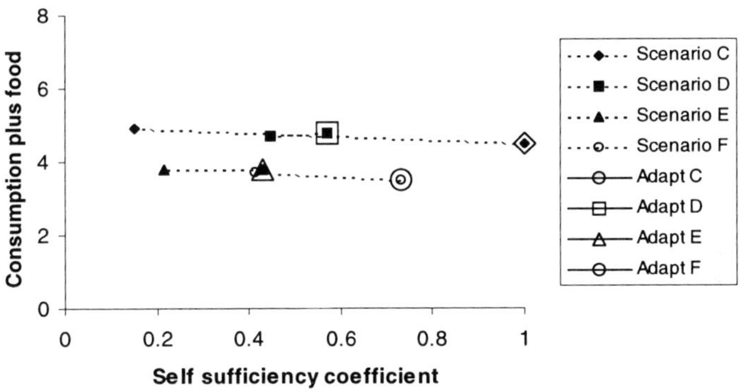

Figure 5; Panel B. Outcomes in 2067 along representative socioeconomic scenarios and climate scenario 3 with the higher population trajectory displayed in terms of an index of consumption goods and food in 2067 (2000 level = 1) and a food self sufficiency coefficient (the proportion of food consumption grown domestically). Smaller points indicate outcomes without adaptation; larger points connected by dotted lines indicate outcomes with adaptation.

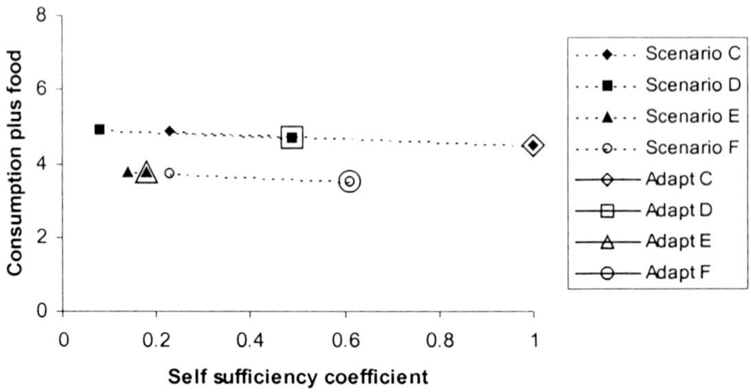

Figure 5; Panel C. Outcomes in 2067 along representative socioeconomic scenarios and climate scenario 6 with the higher population trajectory displayed in terms of an index of consumption goods and food in 2067 (2000 level = 1) and a food self sufficiency coefficient (the proportion of food consumption grown domestically). Smaller points indicate outcomes without adaptation; larger points connected by dotted lines indicate outcomes with adaptation.

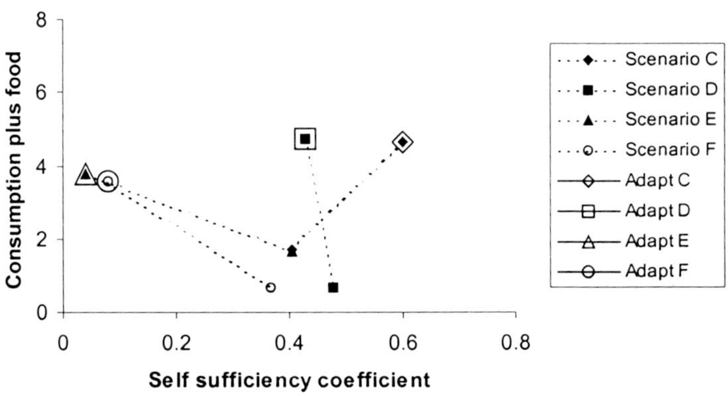

Figure 5; Panel D. Outcomes in 2067 along representative socioeconomic scenarios and climate scenario 9 with the higher population trajectory displayed in terms of an index of consumption goods and food in 2067 (2000 level = 1) and a food sustainability coefficient (the proportion of food consumption grown domestically). Smaller points indicate outcomes without adaptation; larger points connected by dotted lines indicate outcomes with adaptation.

Panel C of Figure 5 meanwhile links outcomes for the same socioeconomic scenarios along the same population trajectory along a more severe climate scenario. Adaptation is again directed, in every case, toward increasing food self-sufficiency; but complete accommodation of the new climate can no longer be achieved. Finally, Panel D focuses attention on the most pessimistic climate future—scenario 9. Here, adaptation was devoted to increasing economic activity. Moreover, food security suffers in three cases, and quite substantially in two where economic robustness is hampered by inefficient investment (scenarios E and F). In fact, only along scenario C, marked by high efficiency in investment and the agricultural sector, could both policy objectives improve with adaptation. The results therefore show that socioeconomic context of the sort suggested by the determinants of adaptive capacity mattered. Indeed, scenarios hampered by inefficient investment displayed diminished capacities to adapt along either the food security or the economic activity scale in both.

The comparisons facilitated by the three panels of Figure 5 certainly support the notion that socioeconomic context can be an important determinant of the capacity to adapt to climate change. Recall that increased food security and not increased economic activity emerged as the goal of adaptation in nearly every case. Scenario by scenario comparisons show that futures hampered by slower economic growth (generally caused by less efficient investment in production capital) portray smaller capacities to adapt irrespective of the severity of climate change even with equally efficient investment in adaptation projects. Adaptation moved the 2067 snapshots of the C and D socioeconomic scenarios (high population growth and high investment efficiency) significantly to the right indicating higher sustainability coefficients for all climate scenarios. Adaptation also moved these points significantly higher along scenario 9, thereby signaling increased consumption as well. Socioeconomic scenarios E and F meanwhile imposed low investment efficiency on high population growth scenarios. Adaptation could still increase food security in all climate scenarios, but much more modestly in most cases. In scenario 9, in fact, adaptation would actually sacrifice food security for increased consumption.

7. Evaluating Adaptive Capacity

What can investigations of this sort into the details of some specific macro-scale adaptations tell us about overall adaptive capacity? Significant insight can be drawn from a process that contemplates the role of each of the determinants identified in Section 1 in supporting the relative efficiency of each adaptation across more qualitative depictions of socioeconomic futures. Yohe and Tol (2002) devised an indexing method designed to organize these thoughts, and this section reviews the structure they propose and the utility of returning to the first principles of adaptive capacity.

7.1 An Indexing Method

Yohe and Tol (2002) constructed an index of the potential contribution of any adaptation option (to be denoted by J) to an indicator of overall coping capacity (denoted by PCCJ) from a step by step evaluation of feasibility factors. These are subjective index numbers that are judged to reflect its strength or weakness *vis a vis* the last seven determinants of adaptive capacity listed above. Values are assigned from a range bounded on the low side by 0 and on the high side by 5 according to systematic consideration of the degree to which each determinant would help or impede its adoption. Let these factors be denoted by ffJ (K) for determinants K = 2, …, 8. An overall feasibility factor for adaptation J can then be reflected by the minimum feasibility factor assigned to any of these determinants; i.e.,

$$FFj \equiv \min\{ffj\ (2),\ \ldots\ ffj\ (8)\}. \tag{1}$$

Each factor inserted into Equation 1 would, in particular, suggest whether the local manifestations of each determinant of adaptive capacity would work to make it more or less likely that adaptation (j) might be adopted. A low feasibility factor near 0 for Determinant K would, for example, indicate a significant shortcoming in the necessary preconditions for implementing adaptation J, and this shortcoming would undercut its feasibility. A high feasibility factor near 5 would indicate the opposite situation; assessors would, in this case, be reasonably secure in their judgment that the preconditions included in Determinant K could and would be satisfied. Note that the structure of Equation 1 makes it clear that high feasibility factors for a limited number of determinants would not be sufficient to conclude that adaptation option J could actually contribute to sustaining or improving an overall coping capacity. The overall feasibility of option J could still be limited by deficiencies in meeting the requirements of other determinants.

The ability of adaptation J to influence a system's exposure or sensitivity can meanwhile be reflected in an efficacy factor EFJ—a subjective index number now assigned from a range from 0 to 1. Efficacy factors thereby reflect the likelihood that adaptation J will perform as expected to influence exposure and/or sensitivity compounded by the likelihood that actual experience would exceed critical thresholds if it were adopted. The potential contribution of any adaptation to a system's social and economic coping capacity can then, finally, be defined as the simple product of its overall feasibility factor and its efficacy factor; i.e.,

$$\text{PCCJ} \equiv \{\text{EFJ}\}\{\text{FFJ}\}. \tag{2}$$

Any method that focuses attention on the feasibility and efficacy of adaptation options through Equations 1 and 2 can be extended to handle the complication of interactions across multiple options designed to mitigate vulnerability to one source of stress or multiple sources of stress. The key here would involve simply keeping track of the degree to which options might complement each other, substitute for each other, or perhaps even work at cross-purposes.

An operational feasibility-based evaluation of an adaptation's potential contribution to overall coping capacity must be informed by a compounding evaluation of feasibility and efficacy for each possible adaptation. The method described in Equations 1 and 2 suggests one way in which this might be accomplished. Note, in particular, that an adaptation option could receive a low coping capacity indicator for many reasons. It might, by virtue of shortcomings in meeting the determinants of adaptive capacity, receive a low overall feasibility index. An adaptation could, as well, receive a low indicator if it were relatively ineffective in reducing exposure and/or sensitivity or if the mechanism by which it could accomplish its tasks were uncertain. Turning now to defining an overall index, suppose that M potential adaptation options had been identified for a specific vulnerability and that they have been assigned potential coping capacity indicators {PCC1, . . . , PCCM}. Yohe and Tol (2002) offer the maximum of the these potentials,

$$\text{CCI} \equiv \max\ \{\text{PCC1}, \ldots, \text{PCCm}\}, \tag{3}$$

as a consistent and workable indicator of overall coping capacity.

7.2 The Application to Egypt Revisited

Returning now to the Egyptian context, Strzepek et al. (2001) highlighted two alternatives that were differentiated in qualitative terms. One, a high capital scenario, would be supported by relatively efficient government investment, attractive investment opportunities for foreign capital and significant domestic investment from the private sector; scenarios C and D depict this possibility. The alternative scenario envisioned lower growth because government investment continued to focus on "mega-projects" that crowded out domestic investment and reduced the attraction to foreign sources of investment; scenarios E and F portray these futures.

Panel A in Table 3 and Panel B in Table 4 offer subjective views of feasibility and efficiency factors for the three adaptations described above. In creating these tables, we judged feasibility factors for each adaptation along two development scenarios. Different readers may assign higher or lower numbers to any or all of them, but it is most important for present purposes to notice that the availability of resources is reduced in the low development scenario for recycling and drip irrigation and not for groundwater drilling. This may appear odd, but a large groundwater project is just the sort of "mega-project" that would appeal to central authorities even along a slow-growth trajectory.

Efficiency factors were meanwhile judged for their ability to handle shortfalls along two climate scenarios in terms of economic activity and food self-sufficiency. Underlying estimates of economic activity supported the GDP estimates, but efficacy in maintaining self-sufficiency can be seen by comparing the post-adaptation outcomes depicted in Panels B and D of Figure 5 with outcomes consistent with the current climate depicted in Panel A. Panel B shows, for example, that self-sufficiency consistent with the current climate could apparently be sustained in 2067 with adaptation along either development scenario. Panel D shows a different story, however. Along the high development scenarios (C and D), modest adaptation could sustain only four-tenths of the 60 to 100 percent self-sufficiency levels that could be achieved if the climate did not change; and groundwater pumping would only increase that fraction to six-tenths. The news is even worse along the low development scenarios (E and F), where only one-tenth of anticipated levels could be maintained.

Combining these factors according to Equations 2 and 3 shows the sensitivity of overall coping capacity to alternative climate and socioeconomic futures; but measures of coping capacity also depend on the

Table 3. Potential capacity indices for adaptation options—high development scenarios, Panel A

Determinant	Recycling municipal water	Drip irrigation	Groundwater drilling
Resources	4	4	5
Institutions	4	4	5
Human capital	4	4	5
Social capital	5	5	5
Risk spreading	4	4	4
Manage information	4	4	4
Public perception	5	5	5
Feasibility factor	4	4	4
Efficacy factor—GDP			
Climate (3)	1	1	1
Climate (9)	0.2	0.2	0.9
Potential coping index—GDP			
Climate (3)	4	4	4
Climate (9)	0.8	0.8	3.6
Efficacy factor—food self-sufficiency			
Climate (3)	1	1	1
Climate (9)	0.4	0.4	0.6
Potential coping index—food self-sufficiency			
Climate (3)	4	4	4
Climate (9)	1.6	1.6	2.4

metric. The overall index for modest climate change along scenario 3 runs from 2 to 4 for low and high development scenarios, respectively, when self-sufficiency is the metric, and it runs from 0.4 through 2.4 for the extreme climate future of scenario 9. The corresponding ranges are 2.8 to 4 and 2.8 to 3.6 for economic activity. Moreover, Tables 3 and 4 make it clear that maximally effective adaptation to severe climate change measured in terms of either metric will require groundwater drilling at some point in the future. The next section explores how planners might be able to discern exactly when to start investing in such a project.

Table 4. Potential capacity indices for adaptation options—low development scenarios, Panel B

Determinant	Recycling municipal water	Drip irrigation	Groundwater drilling
Resources	2	2	5
Institutions	4	4	5
Human capital	3	3	5
Social capital	5	5	5
Risk spreading	4	4	4
Manage information	3	3	4
Public perception	5	5	5
Feasibility factor	2	2	4
Efficacy factor— GDP			
Climate (3)	0.7	0.7	0.7
Climate (9)	0.1	0.1	0.7
Potential coping index—GDP			
Climate (3)	1.4	1.4	2.8
Climate (9)	0.2	0.2	2.8
Efficacy factor—food self-sufficiency			
Climate (3)	1	1	1
Climate (9)	0.1	0.1	0.1
Potential coping index—food self-sufficiency			
Climate (3)	2	2	4
Climate (9)	0.2	0.2	0.4

8. The Value of Climate Information and Expanding Adaptive Capacity

The ability of decision makers to process information was identified in Section 1 as a critical determinant of adaptive capacity. But what would be the value of information about, for example, future climate in this macro context? Computing the value of improved information has been a central theme in much of the economics literature for some time. Indeed, definitions like "the expected value of information is the increase in expected monetary value that would result from the decision maker's obtaining improved information concerning the possible outcomes of his or her decisions" can be found in standard intermediate theory texts (Mansfield and Yohe, 2000, p. 157). Participants in the uncertainty exercises of the Energy Modeling Forum and other researchers have applied this notion to the mitigation and, less widely, to the impacts and adaptation sides of the climate issue; see Manne (1995), Yohe and Wallace (1996), and Yohe (1991) for examples. This section reports results that extend the domain of this line of research to consider the macro-scale adaptations for Egypt described above.

On the adaptation side, of course, the value of information is critically dependent on specific properties of various adaptive responses under consideration. Improved ex ante information can hold negligible value for adaptations that are tied directly to direct observations of current circumstances as the future unfolds as long as the appropriate signal of change can be effectively distinguished from the surrounding noise. Improved information can, however, be enormously valuable for adaptations that involve significant investment in infrastructure when specific thresholds are crossed and/or significant reallocations of resources in anticipation of that investment. Two of the adaptations modeled here, drip irrigation and municipal recycling, fall relatively well into the first category. They would involve modest investment in small projects that could efficiently come on line as climate change produced incremental need. Drilling for groundwater below the Sahara and pumping it to urban or agricultural regions would, in contrast, lie squarely in the second category. Reflecting current expectations of the Egyptian Ministry of Water and Irrigation, the model described above undertakes enormous investment when five-year average flows into Lake Nasser fell 25 percent below current levels. Moreover, the perfect foresight assumed in the Ramsey formulation produces significant reallocations of capital between the agricultural and nonagricultural sectors well in advance of that date.

Returning now to Figure 1 to define a context in which the value of climate information might be significant for this pumping adaptation, note that early portions of not-implausible flow trajectories can be divided into two categories that foretell dramatically different futures. One set, reflected by scenarios 5, 6, 7 and 9, displays significant reductions in flow almost immediately. Scenario 6, the least draconian of the four, crosses the critical 25 percent investment threshold around 2027. All of the other scenarios depict more modest reductions. Indeed, scenario 3, the lowest of these five in the early part of the century, never crosses that threshold. A comparison of these two representatives (i.e., scenarios 3 and 6) can therefore provide some insight into at least the order of magnitude of the value of being able to differentiate perfectly between them.

Computation of this value must, however, reflect the possibility that the government planners responsible for undertaking the investment in pumping infrastructure and directing the requisite prior reallocation of investment between sectors would learn something about their changing climate well before 2027. Faced with the problem of deciding when to do what, planners could, for example, operate under the assumption that something like scenario 3 would emerge until they were convinced otherwise. If scenario 3 did in fact emerge, then their plans could approximate the perfect foresight adaptations described in Section 5. If something like scenario 6 actually materialized, however, then they would have to make a "midcourse adjustment" at some point in time and suffer the cost of not having prepared properly for the future. These costs would climb over time, of course, so planners would feel some urgency to make this adjustment. Unless they waited until they had achieved almost perfect certainty, however, their adjustment decision could still be wrong—an eventuality that would necessitate a second adjustment back to their original plan at the cost of being "off trajectory" for a period of time.

Figure 6 portrays this predicament at some arbitrary point T years into the future. Distributions of five-year averages in flow, based on current interannual variability in flow, are drawn there around two different means. The higher mean corresponds to the expected flow T years from now along scenario 3 while the lower mean reflects expected flow at the same time along scenario 6. The planners' response to this situation T into the future can now be portrayed as one of selecting some critical threshold (the vertical line) and deciding to adjust their plans to accommodate scenario 6 only if the average flow were below this value. The likelihood of a "Type I" error (incorrectly assuming that they were moving along scenario 6 when in fact they were actually experiencing scenario 3) would then be the area under the right-hand distribution to the left of the critical decision threshold. The corresponding likelihood of a "Type II" error (incorrectly assuming that they were moving along scenario 3 when in fact they were actually experiencing scenario 6) would similarly be the area under the left-hand distribution to the right of the critical threshold. Finally, the expected value of having perfect information about the future would be the sum of the present value of the costs of these two types of errors weighted by these relative likelihoods.

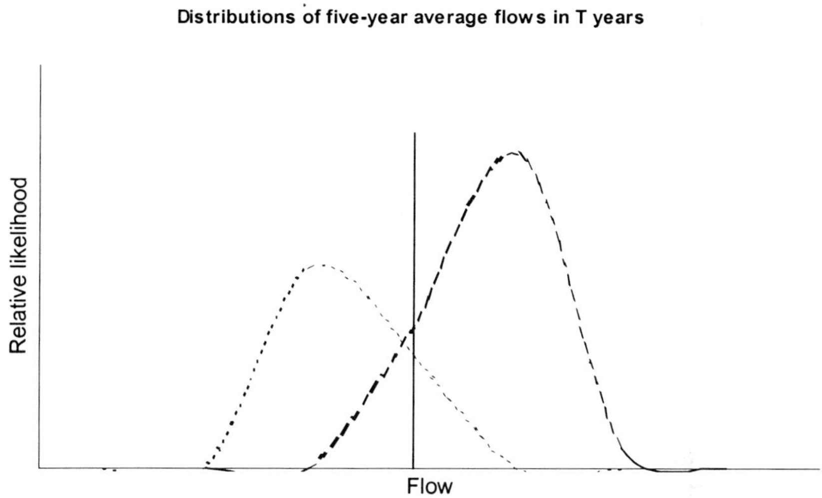

Figure 6. Distributions of five-year moving averages of flow and a critical decision threshold. The upper distribution represents the relative frequency of the five year moving average of river flow, for example, at a particular year (time T) along a given climate scenario. The lower distribution represents the comparable frequencies at time T along a more severe climate scenario. The vertical decision threshold defines two types of errors involved at time T. The likelihood of a Type I error, incorrectly concluding that observed flows below the decision threshold indicate that the more severe climate scenario is emerging, is the area under the upper distribution to the left of the threshold. The likelihood of a Type II error, incorrectly concluding that observed flows above the decision threshold indicate that the less severe climate scenario is still in effect, is the area under the lower distribution to the right of the threshold.

The circumstances depicted in Figure 6 would, of course, change over time. The distance between the means would grow as the future unfolded, and the size of inter-annual variability might also change. The likelihoods of both types of errors could therefore be diminished, but the cost of persisting in the expectation that scenario 3 was unfolding while scenario 6 was actually materializing would surely climb. This is the source of urgency mentioned above; and recognizing it allows the complete adjustment decision to be framed in terms of picking both the year and the critical threshold that minimizes the expected discounted costs of Type I and Type II errors. Moreover, the value of information that would allow planners to differentiate perfectly between scenarios 3 and 6 would precisely equal this minimum.

The economic model described in Section 2 was manipulated to provide time series loss estimates for Type I and Type II errors against scenario 6 under the assumption that planners were operating as if they were experiencing scenario 3. The cost of being incorrect in that assumption was computed in terms of a standard definition of GDP (the sum of consumption plus investment plus exports and minus imports) began to accrue immediately. In contrast, the cost of incorrectly switching to scenario 6 was similarly computed, but it was charged against the economy only when that decision was made. Inasmuch as such a switch would commit the economy to a different investment and planning trajectory for a period of time, those costs were assumed to continue for at least 5 years. At that point, though, planners could either continue with increased confidence on scenario 6 or move back with equal confidence to scenario 3. An interest rate of 5 percent was employed to discount the losses associated with either type of error back to the present.

The flow trajectories depicted in Figure 1 were meanwhile employed to compute the relative likelihoods of these errors on an annual basis beginning in 2005 for an assortment of critical decision thresholds of the sort depicted in Figure 6. Decisions to switch or "stay the course" were based on where those critical values fell on the distributions of five-year moving averages generated under the assumption that the coefficient of variation of interannual variability was constant over time.

The four contours of Panel A of Figure 7 display results of these calculations for socioeconomic scenario C. Scenario 3 is identified there as the basis to signify that these contours were drawn under the assumption that planners would follow the Ramsey prescriptions for scenario 3 until they decided that they should be coping with the manifestations of scenario 6. Each is defined by a threshold expressed in terms of a percentage reduction in the five-year moving average relative to the current flow. Each displays the expected loss associated with applying that threshold at some date in the future—dates that were bounded for each contour by the years during which reductions in the average flow along scenarios 3 and 6 would match the indicated percentage. All of the contours are U-shaped, but that is not a surprise. Expected losses for any threshold should initially fall as the date of

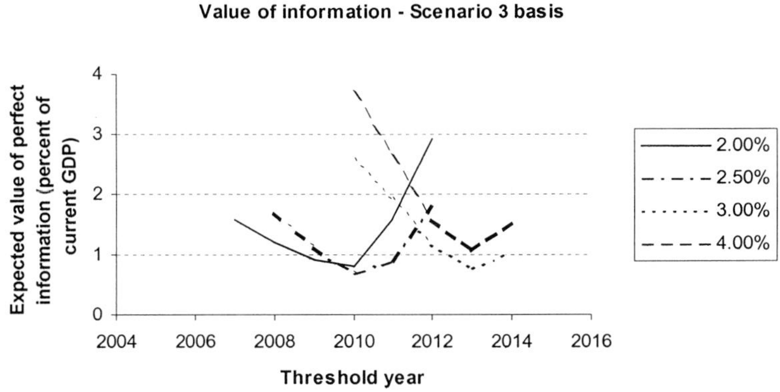

Figure 7; Panel A. Contours indicate the expected discounted loss (measured as a percent of current GDP along socio-economic scenario C) of applying the indicated thresholds in reduced Nile flow to decide whether or not to adjust adaptation from investments that would be most appropriate for climate scenario 3 to those that would be most appropriate for climate scenario 6 for various years in the future.

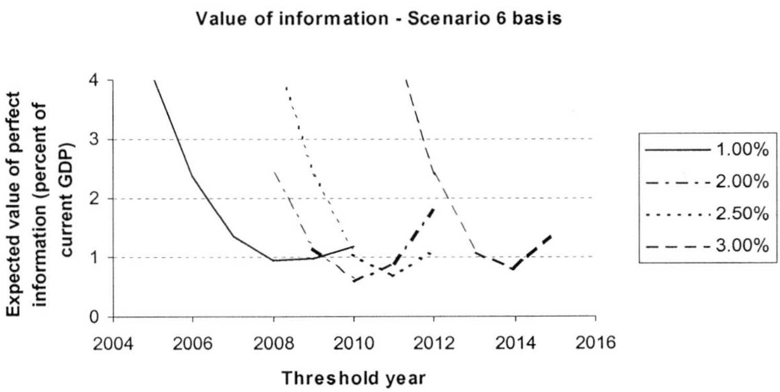

Figure 7; Panel B. Contours indicate the expected discounted loss (measured as a percent of current GDP along socio-economic scenario C) of applying the indicated thresholds in reduced Nile flow to decide whether or not to adjust adaptation from investments that would be most appropriate for climate scenario 6 to those that would be most appropriate for climate scenario 3 for various years in the future.

its application moves farther into the future; but these expected losses would eventually climb when the discounted cost of a Type I error begins to grow relative to the analogous expected cost of a Type II error. Note, finally, that a lower envelope of these contours can easily identify a "global minimum"—a combination of threshold and decision year for which expected discounted losses are as small as possible. Panel A can therefore be read as support for anticipating a midcourse adjustment decision around 2010 using a 2.5 percent reduction as the critical threshold. Doing so would minimize expected discounted costs given a 5 percent discount rate at a value approximating 0.62 percent of current GDP.

Planners could, of course, choose to operate as if scenario 6 were unfolding and force themselves into deciding if a midcourse correction to scenario 3 might be in order. This would reverse Figure 6 and define complementary areas for the relative likelihoods of Type I and Type II errors. It would also reverse the definition of those errors. A Type I error would then involve incorrectly assuming that scenario 3 was being experienced when scenario 6 was real, and a Type II error would involve the opposite. Panel B of Figure 7 displays the threshold contours for this case in which scenario 6 is the basis and the loss calculations were reversed. As expected, the contours portrayed in Panel B are similar but not identical to those depicted in Panel A. Note, in particular, that a "global minimum" of 0.56 percent of current GDP could be anticipating with the application of a 2 percent flow reduction decision threshold against scenario 3 around 2010. The minimum expected loss of planning for the less favorable scenario and correcting when things turn out to be more favorable was smaller, in this case, than expecting good news with a chance of being unpleasantly surprised.

Table 5 displays the results of similar exercises for differentiating climate scenario 3 from 6, 3 from 9 and 6 from 9 for socioeconomic scenarios C, D, E and F. For reference, notice that results for the comparisons 3 versus 6 and 6 versus 3 that just quoted for socioeconomic scenario C are highlighted in italics. All of the other present values of information are measurable fractions of current GDP, and several general trends can be detected. The values recorded above the diagonal are, for example, all lower than the values recorded for corresponding comparisons below the diagonal (the corresponding values are printed with *italics* for 3 versus 6 and 6 versus 3; **bold** for 3 versus 9 and 9 versus 3; and ***bold-italics*** for 6 versus 9 and 9 versus 6). This means that at least one earlier

Table 5. Estimates of the value of information (discounted value as a percentage of current GDP)

Socioeconomic scenario	Actual climate scenario	Assumed baseline climate scenario		
		Scenario 3	Scenario 6	Scenario 9
Scenario C	3	0.00	*0.56*	**0.21**
	6	*0.62*	0.00	***0.31***
	9	**0.26**	***0.32***	0.00
Scenario D	3	0.00	*0.92*	**0.37**
	6	*0.93*	0.00	***0.54***
	9	**0.71**	***1.08***	0.00
Scenario E	3	0.00	*0.24*	**0.15**
	6	*0.24*	0.00	***0.25***
	9	**0.18**	***0.27***	0.00
Scenario F	3	0.00	*0.84*	**0.29**
	6	*0.92*	0.00	***0.51***
	9	**0.30**	***0.57***	0.00

observation is robust: preparing for the worst and (perhaps) adapting to better news always reduced the value of information. Since the value of information was computed as the minimum expected discounted cost of making Type I and/or Type II errors, preparing for the worst seems to be a dominant strategy for all comparisons and all socioeconomic futures. Second, the expected value of information is lower for comparisons of climate scenarios 3 and 9 than it is for comparisons of scenarios 6 and 9. Moreover, it is smaller for comparisons of scenarios 3 and 9 than it is for comparisons of scenarios 3 and 6. These results are, perhaps, surprising. Similar climate scenarios support similar trajectories of investment in anticipation of macro-scale adaptation, so the cost in any given year of preparing for the wrong future should be smaller. It would, however, be more difficult to differentiate between climate scenarios that track each other closely. As a result, more of the future must unfold before a decision to make "midcourse" correction can be made with any degree of certainty; and extra time before a making decision adds extra cost to making a bad one. The values recorded in Tables 3 and 4 therefore make it clear that adding the potential cost of being wrong for a longer period of time consistently outweighs the effect of smaller year-to-year liability.

9. Concluding Remarks

The numerical results reported here are, of course, critically dependent on the model, its specific calibration and its underlying assumptions of perfect rationality. It follows, as usual, that the numbers are not the message. There are instead several qualitative insights that can be drawn from this exercise:

- Adaptation can make a significant difference on this macro scale, especially for pessimistic climate scenarios.
- Socioeconomic context matters in determining adaptive capacity, especially on a macro scale. Scenarios hampered by inefficient investment displayed, for example, diminished capacities to adapt.
- The goals of adaptation at a macro scale can be denominated in terms of objectives other than increased economic activity; food security was the major beneficiary of adaptation to all climates but the most severe in the Egyptian case, for example.
- The value of information that supports early differentiation between two strikingly different climate futures can, nonetheless, be significant even when it is denominated in economic terms. The exercise produced values that could be expressed as measurable fractions of current GDP.
- Early preparation for adaptation can be important because macro-scale adaptations can involve capital reallocation between sectors in anticipation of large future investments in adaptation infrastructure.
- Planning for bad news and adapting to good can be a better choice than the other way around.
- Evaluation of the relative strength of adaptive capacity depends on climate scenarios, socioeconomic contexts and the underlying valuation criteria.
- Working to expand the potential of some options to increase adaptive capacity can create rigidities or cause systems to under-prepare for adopting other options. These omissions can reduce the capacity to cope with more extreme climate futures because the first set of adaptations may be overwhelmed even as more efficacious alternative adaptations become less feasible.
- Experimenting with the value of information calculations showed that:

- Increased variance increases the value of information and delays midcourse corrections; it is even better to plan for the worst in these cases.
- Increased costs increase the value of information, but they seem to be neutral with respect to timing of adjustment decisions.
- Differentiating between climate scenarios that track close to each other in the near term delays the adjustment decision. These delays increase the value of information even though the costs of being wrong in any one year would be relatively smaller.

Even these qualitative conclusions must be evaluated in the context of the aggregate environment from which they were drawn. It is widely known that aggregation masks important information. These estimates of cost denominated either in terms of economic activity or in terms of food self-sufficiency could be too large, because the aggregation missed micro-scale substitution and adaptation opportunities that would be exploited. But they could be too small if aggregation masked constraints at the micro scale that would limit those opportunities. It follows, finally, that these conclusions are best viewed as hypotheses to be explored using a disaggregated dynamic economic model designed to accommodate variance in the adaptive capacities of specific sectors that might be most vulnerable along climate scenarios described not only in terms of flow of the Nile, but also in terms of associated temperature and sea level rise trajectories.

Acknowledgements

We gratefully acknowledge support for this work offered by the Electric Power Research Institute and the National Science Foundation of the United States through its funding of the Center for Integrated Study of the Human Dimensions of Global Change at Carnegie Mellon University through SBR-9521914. We also appreciate insights and comments offered by Joel Smith, by participants at the Climate Impacts and Integrated Assessment Workshop held in Snowmass, Colorado, and by participants in an Economics Department Seminar at Wesleyan University.

References

Barro, R. J. and Sala-i-Martin, X. (1995), *Economic Growth*. McGraw-Hill, New York.

Blanchard, O. J. and Fischer, S. (1989), *Lectures on Macroeconomics*. MIT Press, Cambridge, MA, USA.

Lau, M. I., Pahlke, A. and Rutherford, T. F. (2002), Approximating infinite-horizon models: a primer in dynamic general equilibrium models. *Journal of Economic Dynamics and Control* **26**, 577-609.

Manne, A.S. (1995), A Summary of Poll Results—EMF 14 Subgroup on Analysis for Decisions under Uncertainty. Stanford University, Stanford, CA, USA.

Mansfield, E. and Yohe, G. (2000), Microeconomic Analysis. W.W. Norton and Company, New York.

McCarthy, J., Canziani, O., Leary, N., Dokken, D. and White, K. (2001), *Climate Change 2001: Impacts, Adaptation and Vulnerability*. Intergovernmental Panel on Climate Change. Cambridge University Press, New York.

Strzepek, K., Yates, D., Yohe, G., Tol, R. J. S. and Mader, N. (2001), Constructing "not-implausible" climate and economic scenarios for Egypt. *Integrated Assessment* **2**, 139-157.

Yates, D. and Strzepek, K. (1998), An assessment of integrated climate change impacts on the agricultural economy of Egypt. *Climatic Change* **38**, 261-287.

Yohe, G. (1991), Uncertainty, climate change and the economic value of information: an economic methodology for evaluating the timing and relative efficacy of alternation responses to climate change with application to protecting developed property from greenhouse induced sea level rise. *Policy Sciences* **24**, 245-269.

Yohe, G. and Tol, R. (2002), Indicators for social and economic coping capacity—moving toward a working definition of adaptive capacity. *Global Environmental Change* **12**, 25-40.

Yohe, G. and Wallace, R. (1996), Near term mitigation policy for global change under uncertainty: minimizing the expected cost of meeting unknown concentration thresholds. *Environmental Modeling and Assessment* **1**, 47-57.

Appendix: An Economic Model for Egypt

A nonlinear formulation of the classical Ramsey model developed by Lau et al. (2001) was employed. The primal formulation was based on an explicit representation of the utility function for the single representative household. The social planner maximized the present value of lifetime utility for the representative household.

$$u = \sum_{i=0}^{\infty} \left(\frac{1}{1+\rho} \right)^{t} u(C_t)$$

Where ρ is the time preference rate, C_t is annual consumption in year t, and u(C_t) is the instantaneous utility function.

The representative agent maximized the present value of utility subject to the constraint that output in period *t* was either consumed or invested:

$$C_t + I_t = f_t(K_t)$$

Here, K_t is capital in period t and I_t is period *t* investment. Assuming strict monotonicity and concavity of the underlying production function, it is convenient to think of the production function exhibiting constant returns to scale in capital and a second factor whose supply would be exogenously specified. Labeling the second factor labor, diminishing returns to scale in capital was represented through an underlying production function that exhibited constant returns to scale in labor and capital; i.e.,

$$f_t(K_t) = F(K_t, \overline{L_t})$$

The capital stock in period *t* equaled the capital stock at the start of the previous period less depreciation plus investment in the previous period. Hence, the capital stock was determined by

$$K_t = (1-\delta)K_{t-1} + I_{t-1},\ K_0 = \overline{K_0}\ \ I_t \geq 0$$

where δ is the annual rate of depreciation and the initial capital stock was specified exogenously.

Several major modifications were made to this model to more accurately represent the Egyptian economy. Two production sectors (agriculture and nonagriculture) were represented. Food and consumption goods were both demanded. The economy was opened for balanced trade. Water was added to the production function to provide a link to climate change. And a reflection of Egyptian concerns about self-sufficiency in food production was included in the utility structure. Each of these modifications is described below.

A.1 Two Production Sectors: Agriculture and Nonagriculture

Production was divided into the two major sectors with sector-specific labor and capital. The agricultural sector produced only food for domestic consumption. The nonagricultural sector produced an output that could be invested to produce capital, consumed or exported to pay for imported food.

A.2 Two Demanded Commodities: Food and Consumption

The single household differentiated two types of consumption goods: food and a nonfood consumable. The underlying accounting process noted two types of food. The first sustained a minimum caloric requirement of 2100 cal/per capita/day; it was included as a constraint in the model. The second built supplemental caloric intake up to 1100 cal/per capita/day. Supplemental calories were preferable, on a declining scale, to non-food consumables. This complication added a food balance constraint, and changed the modified objective function in ways that are discussed below.

$$Food\,\mathrm{Im}_t + Agf_t(K_t, L_t, W_t) \geq Foodreq_t + Food\,\mathrm{sup}_t$$

A.3 Open Economy with Balance Trade

The classic Ramsey model does not allow for trade, but Egypt is currently only 70 percent food self-sufficient. Assuming a closed economy would therefore be inconsistent with current and, in all likelihood, future realities. The model was modified to allow nonagricultural output to be exported for equal the amount of food imports subject to the terms of trade for Egyptian nonagricultural output on the world market. The terms of trade parameter, *TOT,* was exogenously specified so that a production/consumption balance equation could be specified:

$$C_t + I_t + TOT * Food\,\mathrm{Im}_t = NonAgf_t(K_t, L_t, W_t)$$

A.4 The Role of Water in Production

Water is an essential factor of production in the fully irrigated Egyptian agricultural sector as well as an important factor in hydroelectricity production, some industries and transportation. Water was therefore included in the model in a dynamic Leontief fashion. Two different by constant rates of technological progress were specified for each of the two sectors so that water use per unit output in either sector would decline over time. Water was also demanded for domestic consumption, and this use received the highest priority. Domestic water use was modeled as a function of income expressed by GDP/capita. Domestic water use typically increases at a very steep rate until incomes reaches approximate $2,000 per capita then it becomes almost constant. Taking these three demands together, a water balance equation of the form was imposed on the model.

$$DOMwat\,([GDP\,/\,cap](t)) + NAGwat_t * NAG(K,L,W)_t + AGwat_t * AG(K,L,W)_t \leq ASWAN_t)$$

A.5 Food Self-sufficiency as Policy Objective

Exogenously specifying target levels of food self-sufficiency frequently resulted in infeasible solutions when the potential impacts of climate change on Nile flows and agricultural water requirements were extreme. To avoid this problem, the level of food self-sufficiency was modeled as a endogenous variable to be determined by the model. Published reports and private communications with the authors reveal that maximizing food self-sufficiency is policy goal of the Egyptian government. To model such a policy, the objective function was modified to reflex a disutility of importing food equal to consumption of nonagricultural good. This was the equivalent of trying to minimize food imports because each unit of food imports caused a double loss in utility—once as a reduction in consumption and then again as a specific negative factor of utility.

7. Do We Have the Adaptive Capacity to Develop and Use the Adaptive Capacity to Adapt?

Ian Burton
Independent Scholar and Consultant
Scientist Emeritus, Meteorological Services of Canada
Toronto, Ontario

Humanity has considerable capacity to adapt to the effects of climate change, but achieving sufficient adaptation presents some serious obstacles. Whatever level of adaptation is achieved is not likely to be distributed equitably unless some important changes are made in international governance and elsewhere. This paper reviews adaptation experience at local, national and global levels, as a basis for exploring future prospects. The issues are explored in the context of the United Nations Framework Convention on Climate Change. A suggestion is made for the development of a global adaptation framework that could help formulate and advance some generally acceptable guiding principles that can simultaneously serve as directives and be used as benchmarks for measuring progress in adaptation.

1. Humanity Can and Will Adapt

Following Chapter 18 of the Working Group II, Intergovernmental Panel on Climate Change (IPCC) Third Assessment Report (Smit and Pilifosova, 2001), the fundamental assumptions of this paper are that humanity can adapt to climate change and that at the maximum level of implementation, adaptation could go a long way toward reducing the impacts of climate change. This does not mean that mitigation or the stabilization of greenhouse gas concentrations is not required. The longer term consequences of anthropogenic climate change, if nothing is done, could be globally catastrophic. The collapse of the West Antarctic Ice Sheet and the shutdown of the North Atlantic Thermohaline Circulation are often quoted as possible examples. The scientific evidence strongly suggests, however, that consequences of such magnitude are many decades or more likely centuries away. Short of global catastrophe the consequences of climate change in the coming decades are still likely to be unacceptably severe, for two reasons. The first is that the global costs of increased climate variability associated with the destabilization of the climate system and the hydrological cycle will be very large. Second, and even more important, the costs seem certain to be distributed in a very unequal way, thus adding one more weight to the burden of global inequality and the causes of global political instability.

The conclusion of this paper is that humanity as a whole will adapt, but that "sufficient" adaptation will be hard to achieve and that whatever level of adaptation is achieved will not be distributed equitably unless some extraordinary steps can be taken, including steps at the international level. It would be a mistake, however, to assume that developing country adaptation is something that can be completely separated from developed country adaptation. We live at a time of the emergence of a global society driven by forces of technology, communications and economic integration. This form of globalization, whatever we may think about it, is a fact. It is increasingly clear that if globalization is to be managed a way in which the undoubted benefits are to be widely shared and a safety net is to be provided for those who suffer adverse consequences from the change, then some advances are required in global governance. Any such advances will have to be negotiated by nation states, which in turn requires an enhanced capacity for collective global leadership. No leaders, however charismatic, can lead in free societies if they get too far ahead of their constituencies. So there are some prerequisites before much progress in more effective global governance will be possible. More people have to become aware and concerned about the global predicament of climate and other adverse environmental changes. The concern has to be expanded to an understanding of the ethical questions involved and what it means to be a citizen of the world and not just another consumer on the planet. The peoples of the wealthier nations have some way to go before they accept a more enlightened view of their self-interest. This in turn depends to some degree on the quality of governance in the developing countries. Governments that fall short of international standards of conduct can easily be dismissed as unqualified to receive assistance or even have their voices heard in world councils.

Adaptation to climate change is by no means the only, or the most important, global problem requiring better governance and more leadership, but it can be seen as pioneering and unequivocally shared territory. If progress cannot be made in the creation of a global strategy for adaptation to climate change, what prospect can there be for the safe navigation of the other serious global problems we face? Unfortunately it is clear from the record to date that there has been too little progress and common understanding. But it is still not too late.

The notion that humanity can adapt to climate change is based on the idea that wealth, science and technology, organizational skills, institutional arrangements, the deployment of skilled people, public information, public support and motivation, commercial and private sector and civil society commitment, and above all political will can be harnessed and sustained over an extended time. Even if all these ingredients are present and used effectively, adaptation to climate change will be a long and continuous process requiring patience and dedication.

It is easy to accept the notion that adaptation is theoretically possible, and even that much can be achieved, but this is different from answering the question, "Why me?" The more pertinent question therefore is whether it can be made to happen in practice. The brief answer of course is, "It depends." This paper explores in some detail just what it is that successful and pervasive adaptation does depend on. The results of this analysis are then used to formulate some suggestions about what is required to strengthen adaptive capacity and the will to adapt. The question in the title of this paper refers to the capacity to act. The paper also raises a more difficult question: "Do we have the will?"

The approach adopted can be formulated into two questions: What can be learned from experience in dealing with climate change? What are the prospects that the lessons of experience can be applied in the coming years? This second question is a means of directing attention to the problems likely to be encountered in mounting a global adaptation strategy.

Both experience and prospects are examined at three levels; the global, the national and the local. "Global" refers primarily to the United Nations Framework Convention on Climate Change (UNFCCC), including the work of the Conferences of the Parties (COPs); the subsidiary bodies, (SUBSTA, SBI); the financial mechanism (Global Environment Facility, GEF) and the implementing agencies, (UN Development Programme, UNDP; UN Environment Programme, UNEP; and the World Bank); the Kyoto Protocol mechanisms; and the IPCC. There are many others including the regional banks, other UN special agencies and international nongovernmental organizations, both of the civil society and private sector persuasions.

"National" refers, of course, to the policies and actions of national governments, in both their internal role and the perception and projection of their domestic interests at the international level. "Local" refers to subnational organization, especially at the "grass roots" or community level, including households, families and individuals.

The structure of the paper therefore follows the 2 x 3 matrix shown in Figure 1.

It has often been argued or assumed that because anthropogenic climate change is a new phenomenon the subject cannot be addressed empirically. This is largely the case for the atmospheric science of climate change, and so general circulation models (GCMs) and climate scenarios have been used to fill the gap. Empiricism is excluded from adaptation science only if one clings to the fiction that a clear and distinct boundary can be established between climate change and current climate variability. The fact that this fiction continues to be seriously entertained stems from the architecture (perhaps inevitable, but nevertheless bizarre) of international agreements that have been interpreted to mean that assistance will be available under the Framework Convention for adaptation to climate change, but not for adaptation to "normal" climate, since adaptation to normal climate is

	Adaptation experience	**Adaptation prospects**
Local	Increasing losses: need to adopt better measures	Strengthen empowerment and adaptation capacity
National	Increasing losses: need for adaptation policy	Link climate change Adaptation with disaster mitigation in development policy
Global cooperation	Weak governance and leadership, growing inequity	A common framework would advance an agreement

Figure 1. Structure of the paper

assumed to be taken care of in normal development assistance. In fact as the growth in disaster losses demonstrates all too clearly, disaster mitigation has not been a great success even in the absence of climate change and in spite of the recently completed International Decade for Natural Disaster Reduction.

Once this set of blinkers is removed it becomes clear that there is an enormous wealth of human experience in responding or adapting to climate variability and extremes as well as longer term, nonanthropogenic climate change. The former kind of experience has been studied and reported in the natural hazards and disaster literature and elsewhere. The latter has been explored extensively in the "climate and civilization" genre of studies. Economic and social decline, mass migrations and even the collapse of ancient civilizations have been attributed to changing climate. A close reading of this literature invariably reveals that other factors besides climate change and its consequences are at work, and it can be observed that the more distant in time, and the less the understanding of social, economic and political phenomena (because the evidence has been lost), the greater the weight given to climatic factors. Apparently past climates can be reconstructed rather more readily than past socioeconomic and political systems.

For the purposes of this paper, the empirical review is limited to more recent experience with climate variability and extremes. Before leaving "climate and civilization" to the history books, however, a few important observations can be made about which it may be useful to remind ourselves. This is admittedly something of a digression from my main argument and the less patient reader may wish to skip over the next section.

2. The Adaptive Species

That human beings are the most adaptive species on the planet is a basic component of our collective self-image. Human adaptability is the reason given in standard textbooks in archaeology and anthropology to explain the success of the human species in spreading out from presumed East African origins to occupy and thrive in virtually all the climatic environments on the earth. To some extent this adaptability is physiological and evolutionary in a neo-Darwinian sense. Perspiration and shivering are short-term physiological responses, and in the longer term human bodies can acclimatize in various ways. Intelligence is, however, the much more powerful basis for adaptation, and it is on intelligence that future adaptation will be based. This thought has led my colleague at the University of Toronto, Tad Homer Dixon, to write of "the ingenuity gap". "Does humanity", he asks, "have enough ingenuity to cope with the problems of our own making?" (Homer Dixon, 2000).

Certainly the arrival on the public and international agenda in the 1990s of anthropogenic climate change has quite suddenly challenged our self-confident assumption about our own adaptability. It suddenly appears important that disease organisms and vectors also have great adaptability in a changing climate. So we are beginning to recognize that a new order of adaptation is now required which will have to be faster and more effective because anthropogenic climate change will be much more rapid than historical changes in climate and more rapid than the past migrations of peoples from one climatic region to another. For this reason the new adaptation will be more costly, and will require new or advanced technology, higher levels of skill and training and greater organizational and institutional capacity. There is little doubt that the resources, skills and technology can be found. But since these requirements can be met more readily in some countries than others, it is argued that some countries have a deficit in adaptive capacity. This has given rise to the recognition of a need to develop international capacity to strengthen national and local adaptive capacity. The "ingenuity gap" therefore exists more in the area of collective international understanding and cooperation. How can this challenge be met?

The need for greater adaptive capacity, in this sense, has not suddenly appeared in response to climate change. There is already a considerable lack of adaptive capacity. Climate change comes as an additional threat to a world that is currently seeking ways to cope with gross inequities in wealth and power. While the inequities are not themselves new (the Romans were civilized and rich when the inhabitants of the British Isles were not, and the technology of medieval Europe lagged well behind that of China), the shrinking of the world into a global village is new and today's stark inequities are not sustainable or compatible with a peaceful and orderly world functioning under the rule of law. Stronger adaptive capacity in developing countries can help to reduce poverty and vulnerability and promote economic growth and development, and at the same time reduce inequities. This would bring global benefits of considerable magnitude. The achievement of such desirable objectives also requires adaptation in the developed countries. There are three aspects to this. First, developed countries will also have to adapt to climate change. So far most developed countries have tended to assume that they already have the capacity to adapt when and where necessary and have been slow to take on board the complexities and advance planning that are required. Second, in adapting to climate change developed countries can best do so in a manner that does not further aggravate the problems of developing countries. The third and biggest adaptation step in developed countries, which also includes or contains the previous two steps, is coming to terms with the social, political and governance implications of globalization.

While the necessity to strengthen adaptive capacity in all countries is, or should be, self evident, it should not be based on the assumption that no adaptive capacity exists even in the poorest countries and communities. On the contrary any program to strengthen adaptive capacity should be built upon what already exists and is found in our genetic as well as cultural inheritance. It can be agreed that in the face of anthropogenic climate change the world as a whole needs more adaptive capacity and that since the problem is global both in cause and consequence, the capacity for adaptation that exists should be shared and made widely available. Steps are therefore required to build more adaptive capacity in all countries and to ensure that it is more equitably distributed. This involves the sharing and transfer of technologies (both hard and soft) for adaptation.

At the same time it should be recognized that much of the responsibility for adaptation to climate change (always including variability and extremes) rests at the local, community and individual levels. Nothing in the argument for national and international action should be interpreted as a reason for people to do less for themselves or to encourage economic or psychological dependency. Adaptation is something we do for ourselves, and while national and international policies can help they are of little use without local motivation and empowerment.

3. The Context of the Framework Convention

By tracing the evolution of ideas about adaptation and adaptive capacity from the negotiation of the UNFCCC in 1991-1992 to the present, it is possible to identify where some of the obstacles and constraints are to be found. Removing these obstacles and constraints requires policy evolution and initiatives. The creation of the necessary confidence and understanding to adopt policy innovations requires more knowledge to be brought to bear on the issues. This in turn means more original research. But more important than original research is the need to assess existing knowledge from other policy domains and to bring it to bear on adaptation and adaptive capacity.

Chapter 18 of the IPCC Third Assessment Report (Smit and Pilifosova, 2001) draws attention to the need to adapt to climate change and variability, including the need to take account of present vulnerability. The report concludes that the poorer developing countries in the tropical regions are the most vulnerable to climate change because they have the least capacity to adapt. General adaptive capacity is said to be a function of such factors as wealth, access to technology, skilled human resources, effective organization and institutions and the like. It is also a function of exposure to the effects of climate change: small island states, countries with dense population in low-lying coastal zones and those in regions of high atmospheric hazards such as frequent tropical cyclones or low and uncertain rainfall levels are considered to be most vulnerable and most in need of adaptation. So too are countries where the economy relies heavily on climate dependent activities such as agriculture and tourism.

Conversely the richer industrialized nations are assumed to have greater capacity to adapt. Although the actual change in climate as suggested by change in mean temperature is expected to be greater away from tropical regions, middle-latitude countries are assumed to be less vulnerable. A conclusion that can be drawn from this analysis is that adaptive capacity is more important than the amount of actual climate change in determining impacts.

The general (although not universal) acceptance of this diagnosis has led to strong demands from developing countries for financial and technical assistance to help meet the costs of adaptation. This demand is supported by the ethical argument that since the northern industrialized countries are historically responsible for the vast majority of greenhouse gas emissions, they bear a disproportionate share of the responsibility for climate change and its impacts on developing countries. The developed countries accepted this argument in principle in the negotiations that led to the UNFCCC. Article 4, Clause 4 of the Convention states that "the developed country Parties . . . shall also assist the developing country Parties that are particularly vulnerable to the adverse effects of climate change in meeting the costs of adaptation to those adverse effects".

The principle of assistance was clearly established, but the record of the years since the Convention came into force (21 March 1994) has been marked more by vigorous debate about the appropriate way to implement Article 4, Clause 4, than by sustained action.

At the first Conference of the Parties (COP 1, held in Berlin in 1995) a decision was made to proceed with adaptation in a series of stages (Decision 11/CP.1). The text of the decision reads in part as follows:

> 1(d) Regarding adaptation, the following policies, programme priorities, and eligibility criteria should apply:
> Adaptation to the adverse effects of climate change, as defined by the Convention, will require short, medium and long term strategies which should be cost effective, take into account important socio-economic implications, and should be implemented on a stage-by-stage basis in developing countries that are Parties to the Convention. In the short term the following stage is envisaged:
> Stage I: Planning, which includes studies of possible impacts of climate change, to identify particularly vulnerable countries or regions and policy options for adaptation and appropriate capacity-building.
> In the medium and long term, the following stages are envisaged for the particularly vulnerable countries or regions identified in Stage I:
> Stage II: Measures, including further capacity building, which may be taken to prepare for adaptation, as envisaged by Article 4.1 (e);
> Stage III: Measures to facilitate adequate adaptation, including insurance, and other adaptation measures as envisaged by Article 4.1 (b) and 4.4.

Three years later at the fourth Conference of the Parties, in the Buenos Aires Plan of Action (COP 4, held in Buenos Aires in 1998), it was decided to proceed to Stage II activities and "in particularly vulnerable countries and regions identified in Stage I activities, and especially in countries vulnerable to climate-related natural disasters, taking into account their preparatory adaptation planning frameworks in priority sectors, the completion of Stage I activities, and in the context of their national communications" (Decision 2/CP.4 1(a)).

Two observations can be made immediately about these negotiated decisions. First, they are extremely cautiously worded. Here is no evidence of a rush to adaptation for the benefits it may bring. What is mostly proposed is planning and studies to identify vulnerabilities and policy options. Even in

Stage II, which was initially envisaged as a "medium to long term" activity, the measures allowed include only capacity building and preparations for adaptation. No actual adaptation is apparently expected. In Stage III the purpose of the measures is described rather ambiguously as "to facilitate adequate adaptation". A dispassionate observer could be forgiven for concluding that the text reflects the sort of creative ambiguity that sometimes passes for agreement in international negotiations.

The second observation, which supports the first, is that very little has been done to implement the decisions. Relatively small sums have been made available through the GEF (the financial mechanism of the Convention). A very small number of studies has been supported (in Bangladesh, the Pacific Islands and the Caribbean), and the total expenditure for adaptation has been very small compared with that for mitigation over the same period. In addition small sums have been made available to support the preparation of national communications through the UNDP-GEF National Communications Support Programme, and some programs of study of impacts have been sponsored including the US, the UNEP and the Dutch Country Studies programs.

Why was so little attention given to adaptation in all the climate debates and negotiations of the 1990s? First, the climate change issue was scientifically and socially constructed as a classical pollution problem to be addressed by reducing emissions, and there was a corresponding failure to recognize the potentially powerful role of adaptation. This perspective was supported by the dominant role of atmospheric science in putting the issue on the agenda and having little interest in adaptation. It was also supported by environmentalists who saw in the climate change issue an opportunity to advance other parts of the green agenda in the areas of biodiversity, the protection of natural ecosystems, the reliance on fossil fuels and nuclear power rather than renewable energy, and that *bete noire* of the greens, the private automobile, among others. Many of the developing countries were also happy to focus attention on emission reductions, perhaps because the prospect of the transfer of funds and technology for mitigation seemed to offer much greater prospects than did those for adaptation, and perhaps also for its symbolic, ethical and political value as an argument to use in negotiations with the richer countries. Even the private sector interests most on the defensive, including the fossil fuel and automobile industries, did not speak up for adaptation. Perhaps they preferred to deny or trivialize the problem, and to speak favorably of adaptation would have been a tacit admission of the existence and seriousness of climate change.

A second reason for the lack of attention has to do with the distribution of benefits of mitigation and adaptation. Climate change was identified as one of a new class of environmental problems that are truly global. All countries are involved, although to a differing extent, in both creating the problem through emissions and suffering the consequences in the form of adverse impacts. For this reason the problem was carefully separated from other environmental and development problems that could be thought of as nonglobal. Recognizing that there would be unequal costs and benefits associated with global environmental problems, agreement was reached to establish the GEF as a financial mechanism, on the understanding that the services of the GEF would be limited to global environmental problems. Mitigation and adaptation were perceived, incorrectly in my view, to fall on opposite sides of this divide. Mitigation of greenhouse gas emissions would yield global benefits, but the benefits of adaptation would fall overwhelmingly in the place where the adaptation measures were taken. Many of the early suggestions for adaptation fell foul of the assertion that there would be no global benefits. This holds true only if a very restricted, climate-centered view of benefits is adopted. The implication, surely not believed by any of the players, is that actions such as the protection of coastal lands, avoidance of climate "refugee" migrations, poverty reduction and prevention of social and political instability do not yield global benefits.

Third, the argument about global and nonglobal benefits suited a prevalent policy perspective among some members of the donor community that development assistance had not worked well enough (the so-called "donor fatigue" effect) and that greater reliance should be placed on private sector investment to accelerate development. Moreover the argument for development assistance for climate change adaptation was perceived as an additional and unwelcome burden by overworked development task managers in aid and financial institutions who have been asked to incorporate into their already complex responsibilities environment, gender, poverty and other issues that they see as peripheral to their central economic objectives.

There is yet one more explanation for the relative neglect or rejection of adaptation throughout the 1990s. While the distinction between global and nonglobal environmental problems was seen as important for policy reasons, in the climate impacts adaptation field there is no scientific basis for making such a distinction. How much of the present variability of climate, and how much of the longer term trends can be attributed to anthropogenic climate change, remains uncertain and is likely to continue to do so. While there is a growing popular perception that climate extremes are being affected by climate change, the most that climate science can confirm is that recently observed changes in climate variability and extremes are consistent with the sort of changes that might be expected from, but cannot be definitively attributed to, climate change. If the theoretical distinction between anthropogenic climate changes and "normal" climate variability is dropped, then there is almost no limit to the claims that can be made for assistance to meet the costs of policies and measures designed to adapt to climate change and reduce vulnerability to climate variability. Thus at a time of declining aid budgets both bilateral and multilateral development assistance programs came to fear being exposed to a "black hole" of demands for adaptation assistance, and the potential divergence of limited funds into climate change adaptation and away from higher priority needs.

4. Adaptation Experience—National

According to the available statistics, losses from climate-related and other natural disasters have been growing rapidly in real terms in recent decades in both developed and developing countries. Data from Munich Reinsurance, the International Red Cross, and other sources show an accelerating increase in both insured and economic (noninsured) losses.

There are many deficiencies in these data on losses, and the conclusions based on them should be treated with caution. Careful analysis has been made of the data on property losses from floods and hurricanes in the United States since 1926, when national estimates of the amount of property at risk (fixed tangible reproducible assets) first became available. This analysis shows a clear indication of a rise in losses in constant dollar terms, but no significant growth is observed as a percentage of material wealth (Goklany, 1999, 2000). Extrapolating from this study, and other less rigorous information, it may be inferred that in developed countries the growth of losses is largely a reflection of increases in national wealth.

Recent experience in developing countries suggests that the same conclusion may not hold. Certainly loss estimates from some well-reported recent climate-related disaster events such as Hurricane Mitch in Central

America, the Orissa tropical cyclone and floods in Mozambique amount to much higher proportions of gross domestic product than is common in developed countries. In the matter of a few days, losses equivalent to the value of a decade of economic growth have been recorded.

The story with respect to mortality and morbidity is much more encouraging. Loss of life from natural climatic hazards decreased dramatically over the 20th century in the developed countries, and a similar trend can now be observed in developing countries.

What explanations lie behind these observations? While it may be comforting that in developed countries losses are more or less a constant proportion of national wealth, this is an anomaly during a period of rapid advance in science and technology. Risks from natural hazards are much better understood. High risks areas are precisely mapped, forecasting and warning systems are much more timely and reliable, and the technology of building materials and construction have advanced to the point at which much more hazard-resistant structures could be built. A large part of the explanation for growing property losses and declining mortality seems to be that we are much better at emergency response (forecasts, warnings and evacuation) than we are at the longer term applications needed in building construction and spatial planning and zoning. This has led in the United States at least, and to some extent in other developed countries as well, to a shift in emphasis from emergency preparedness and response to disaster mitigation. This latter phrase is used in the hazards community to mean something very similar to the term adaptation in the climate change literature.

The new advocacy of mitigation (risk reduction) in the hazards policy arena has yet to demonstrate results. While rapid results are not to be expected, there are also reasons to doubt its likely effectiveness. These mostly have to do with local and community response (White et al., 2001).

The developing countries have experienced some time lag in the application of new scientific understandings of natural climate-related hazards. There are signs of considerable success in the reduction of mortality as forecasting has improved, warnings are more effectively disseminated and heeded and transport services and mobility improve. In many countries however there is no policy of mitigation, or where it exists on paper it is not applied in practice. Rather the processes of development are such that vulnerability is being increased by the expansion of human settlements and resource use into high hazard areas on the coast, in flood plains, on steep hill slopes and into areas of low and unreliable rainfall. It is not just a matter of the location of human settlements and economic activity. The quality of construction is often low and building codes are not followed or enforced. The magnitude and frequency of the hazard events themselves are also being increased by environmental degradation such as deforestation, soil erosion, offshore mining of sand, destruction of coral reefs and the like (Abramovitz, 2001).

5. Adaptation Experience—Local and Community

Many of the shortcomings of national policy can be explained in terms of local, community and individual circumstances. In the developed countries, again taking the United States as the example, there are strong forces that serve to encourage risk taking and growth in exposure to hazards. Building codes and land use restrictions are not popular with developers, and are frequently avoided or evaded. Those responsible for construction are often not the long-term owners: properties are built and quickly sold so that the risk to the builder is short term. Buyers may be unaware of the risks they are taking, or may discount them as low probability events not likely to occur in the financial lifetime of an investment. Risk taking may be further encouraged by the amenity value of properties in coastal areas, along rivers or in areas subject to fire risk. Such risk taking in search of amenity value is often a side effect of wealth. Whatever the reasons for risk taking, they can be further encouraged by the provision of insurance or the knowledge that in the event of a disaster more or less generous public assistance is likely to be available.

Thus property losses mount, imposing extra costs on the public purse and threatening the insurance industry, but apparently falling within the range of socially acceptable risk because loss of life is small, and the level of economic disruption is not a serious threat and often does indeed bring regional benefits by stimulating the business of repairs and reconstruction.

A recently completed assessment of natural hazards in the United States is entitled *Disasters by Design* (Mileti, 1999). It concludes that losses will continue to increase at least in line with the growth in national wealth unless a major "culture shift" can be socially engineered comparable with the recent shift in social norms and expectations with respect to diet and exercise, smoking, seat belt use, the "war on crime" and the practicing of safe sex.

While not 100 percent successful, it does appear that such campaigns can be effective in time, making allowance for the progressive increase in awareness, understanding and eventual compliance. It may be possible in the United States, given the political will, to launch a social campaign that will, in time, separate the trend of losses from natural hazards from the growth in national wealth.

In the developing countries both losses and vulnerability increase more as a result of poverty rather than wealth. Low quality construction and choice of exposed locations often stem from lack of choice. Many dispossessed or poor rural populations migrating to urban areas in search of economic opportunity are forced into unorganized squatter settlements in hazard prone locations, where basic necessities such as safe water supply and sanitation are lacking and where it is not practicable to insist on standards of construction. This sort of constraint also applies to public infrastructure, where higher design standards mean higher costs to avert what may be perceived as relatively low risk compared with the immediate benefits of a functioning road, bridge or communications system. When the inevitable disasters occur the financial resources to absorb the losses are not available, either locally or nationally.

6. Adaptation Experience—International

International response to natural hazards and disasters has been largely directed to emergency relief and rehabilitation. Up to the 1970s this was an *ad hoc* process at the discretion of whichever nations were able and willing to provide assistance. In addition to national governments, many civil society organizations such as the Red Cross provided emergency relief. To help coordinate this process the United Nations created the Office of the Disaster Relief Coordinator (UNDRO). The story of UNDRO and its successors has not been a particularly successful one. National governments did not wish their efforts to be coordinated and even civil society organizations did not accept the idea. Coordination, still less direction by the UN, detracts from their own visibility. Despite this reluctance the Department of Humanitarian Affairs persisted in attempts to obtain and allocate specific donor tied and designated relief funds. The benefits of such activities have been small. A secondary role has been the provision of scientific and technical information and the promotion of training courses and workshops. There may have been significant benefits from such activities but they are hard to measure or even identify. No independent post-audit evaluations seem to have been carried out.

Faced with the rising toll of losses in both the developed and the developing countries, in the mid-1980s scientists in the United States promoted the idea of an international decade. This took shape in the form of the UN Decade for Natural Disaster Reduction 1991-1999, and a secretariat for the Decade was established in Geneva. The initial concept was science driven. The idea was that the world-wide dissemination and application of scientific and technical knowledge about the magnitude and frequency of hazard events, the locations of their potential occurrence and the use of newly available improved forecasts and warnings could help to significantly reduce vulnerability and contain losses. This wisdom was challenged in the early years of the decade and the mid-decade conference in Yokohama in 1994 made a major course correction toward the goal of mitigation using a much wider array of approaches. This change came late, and the direction of the Decade was hard to change. By the end of the decade the most favorable consensus to be found was that UN Decade for Natural Disaster Reduction had been a very modest success. Many had much more critical views. From the outset few developed country governments had shown much interest. Very little in the way of additional funding was forthcoming, and while national committees for the decade were established on paper in many developing countries few of them were comprehensive in their professional composition and most were not very active. As the Decade came to an end many argued for some successor organization or activity, and an interagency battle took place between OCHA, the Office for the Coordination of Humanitarian Affairs, which regards disasters as a humanitarian problem and therefore within its legitimate territory, and UNDP, which argued that given the new focus on mitigation (read adaptation) the reduction of disasters was a development issue. OCHA won, and with some limitations on its ownership, an International Strategy for Disaster Reduction was established with an interagency working group and a small secretariat in Geneva, with virtually no operating fund except that required for its own payroll. It occupies its time preparing technical reports and exhorting governments to action, mostly in vain. The split in the disaster community remains as deep as ever. Some would keep the humanitarian focus and strengthen the links with other disaster response, including civil conflict, genocide and the results of terrorist attacks. Others would redirect the effort toward mitigation/adaptation as part of the development process and make common cause with the climate change community. Both an "all emergencies" response capability and a "hazard mitigation" approach to development are needed at the international level. Unfortunately looking sideways often slows or prevents forward momentum. Other agencies, including the World Bank, continue to develop their own disaster programs. There is much talk of better integration and coordination, and attempts have been made such as the World Bank's "ProVention Consortium".

Although the division between humanitarian disaster assistance and disaster mitigation has often been cast as an interagency issue, at a more fundamental level it is a functional issue. No agency charged with emergency humanitarian response can expect to devote sufficient resources and amass the necessary skills to deal effectively in the development aspects of disasters. Humanitarian relief and hazard risk reduction are like oil and water. Only when disaster mitigation (risk reduction) is fully incorporated into the thinking and practice of development can real progress be expected (Burton and van Aalst, 1999).

The integration or harmonization of climate change adaptation and disaster mitigation activities in a development context is only one of a number of similar opportunities. So far climate change adaptation has been promoted almost entirely independently from the work in biodiversity protection and the efforts to combat desertification and land degradation. These and other multilateral environmental agreements should be brought closer together.

The emergence of climate change adaptation onto the global agenda in the recent past has given some renewed hope to those in the development agencies who would like to see policies and measures to reduce vulnerability to climate variability and extreme events built into national economic development. What are the realistic prospects that this could happen?

7. Adaptation Prospects—National

In the developed countries much is likely to depend on the success of the move to strengthen the new focus on the mitigation of natural hazards. This is already being helped by the interest in climate change adaptation, and if the two policy domains can be more closely integrated so much the better. There are reasonable prospects that the damages from climate variability, even with climate change, can be contained as a fairly constant proportion of national wealth. This recognizes that losses will nevertheless continue to rise, and that climate-related disasters will continue to occur as an unavoidable consequence of population and economic growth and private risk taking.

The remaining challenge is to thus to continue to promote mitigation in all its forms as vigorously as possible, to minimize loss and disruption when the inevitable does occur and to provide compensation for losses through a combination of public relief and private insurance without encouraging more ill-advised risk taking.

In the developing countries the potential convergence of hazard mitigation with climate change adaptation holds out some real promise of progress. On the other hand the relative lack of success in the developed countries in controlling the rise of disaster losses suggests something of the major challenge to be faced. Improved forecasts and warnings are already bringing down the level of mortality and morbidity, but rapid and continuing improvements in infrastructure design and location, and in management practices in the renewable natural resources sectors, including water, will be difficult to achieve in the circumstances of rapid population and urban growth unless there is also rapid economic growth with sufficient built-in attention to environmental and social sustainability.

In these circumstances the potential availability of additional funds to assist with adaptation to climate changes comes not a moment too soon. If the experience of developed countries is any guide, the funds can easily be wasted or can make matters worse. Effective use, at a minimum, requires the integration of climate change adaptation into national development policy, not simply at the macroeconomic level, but also in sectoral and regional policies and in building codes and physical planning. A recent informal survey indicates that many developing countries recognize the need, and are beginning to consider how to proceed. A danger is that funds will be misallocated to projects and programs that are short term, politically expedient or, worse still, serve to facilitate more development of a kind that will increase rather than reduce vulnerability, in a word, maladaptation.

8. Adaptation Prospects—Local and Community

Much adaptation activity takes place at the local level. National policy sets the context, but what actually happens on the ground can and should depend on the local communities. There is a well-recognized problem of a top-down development process in which well-meaning efforts are made by outside actors, but collapse when the outside interest move away. Development assistance has received, and deserves, much criticism for the practice of short-term, externally driven projects. The current reformist mantra is "stakeholder involvement", coupled with the strengthening of adaptive capacity at the local level. The sentiments are good, but the actual delivery is a very different matter.

Two of the major issues are local control and empowerment, and the size and delivery of the funding package. In many developing countries the capacity of local government is weak. Sometimes what power exists is highly concentrated in a few hands. Local leadership may not be very representative of the community, and may have vested interests that do not coincide with priority local needs. Thus the problem of stakeholder involvement presents a challenge to local power structures, and to the external drivers. To connect properly and in a sustained manner with "stakeholders" is a highly labor intensive activity. It can add significantly to the transaction costs of a project, and the length of time required from conception to implementation. In contrast the external funding organizations are concerned about reducing transaction costs and streamlining development assistance.

The prospects for adaptation at the local and community level thus depend to a very large degree on the local circumstances, on the political culture of the region or country and on the relationships that exist between national and local levels of government. Strengthening adaptation capacity for climate change at the local and community level is highly dependent on context.

9. Adaptation Prospects—International

The prospects for international action to create capacity for climate change adaptation have improved dramatically in the last 2 or 3 years, most clearly in the case of the COP 6 decisions made in Bonn in July 2001. So far these are decisions in principle and much will depend on how they are implemented.

Four questions seem most relevant. Can adaptation be adequately marketed to achieve public awareness, understanding and conviction? Can adaptation assistance be adequately coordinated? Will adaptation assistance be effectively used by integrating it into national development? How can priorities be set among the many competing demands and claims to be among the "particularly vulnerable"?

One result of the sudden coming into fashion of adaptation is that donor funds are likely to be made available in a variety of ways through a variety of sources. Two new funds were created in the Marrakech Accords. The Special Climate Fund (voluntary donor contributions) has a broad mandate that includes adaptation. The Least Developed Countries Fund (voluntary donor contributions) has already received some contributions to support the preparation of National Adaptation Plans of Action (NAPAs) in the 46 least developed countries. These funds are in addition to the adaptation money that is expected to come into being by means of a levy to be imposed on transactions under the Clean Development Mechanism of the Kyoto Protocol. For these funds the developed countries prefer GEF as the financial mechanism, but it has been sharply criticized by the non-Annex 1 countries. GEF has three implementing agencies (UNDP, UNEP and the World Bank). There is a logical division of labor between these agencies. UNDP is well placed to provide policy and technical pre-investment assistance of the kind implied in the three adaptation stages listed above in section 3. The main strength of UNEP is its specialized science-based work complementary to that of the World Meteorological Organization (WMO). It might be presumed that the World Bank is all about investments, but it has recently broadened its horizons to become a "knowledge bank" focusing on poverty reduction. These core mandates are more easily listed than defined, and inevitably they tend to expand into newly defined problem areas. In the competitive world of interagency rivalry the danger is that the areas of overlap become areas of conflict rather than cooperation.

Not all adaptation assistance will be channeled through GEF. The regional banks are developing their own programs. Some bilateral donors are already supporting adaptation projects as part of their regular assistance activities. The Red Cross and Red Crescent, the Stockholm Environment Institute, the International Union for the Conservation of Nature and Natural Resources (IUCN), the International Institute for Sustainable Development

(IISD), the International Institute for Environment and Development (IIED) and other civil society organizations are moving in the same direction. And as the actual and potential impacts of climate change become more clearly defined and understood the private sector will do much of its own adapting. Multinational corporations can be expected to take climate change into account in their own operations and are likely to become one of the main actors, especially in the protection of their own investments and in the search for the new business opportunities that climate change will bring.

Close coordination or integration of all these diverse adaptation activities is neither possible nor in the end desirable. On the other hand some global architecture for cooperation is clearly required. Things will go better if information is shared, if a reasonable division of labor can be clarified and accepted and if a common intellectual and policy framework can be built to help guide the effort (Burton and Lim, 2001; Burton et al., 2002).

One important function of a common framework would be to ensure that adaptation funds, from whatever source and by whatever channel they are delivered, are integrated into national development policies and plans. The UNDP-GEF National Communications Support Programme has recently taken an initiative in this direction, and its further development with the cooperation of other organizations is now proceeding.

The question of priorities has been hanging in the background for some time, and is now likely to come more to the fore. The international donor community is strong in suggesting that priorities need to be established within countries, but the issue of priorities between countries (except for small island countries and the least developed) has been largely avoided, at least in public debate. There has been some work on creating indices of vulnerability to climate change at the national level. These can provide useful background to decision-making and negotiations, but should not be, and are not likely to be, used as a formula for allocating adaptation funds.

Probably what will and should emerge is a process guided by the best available knowledge on absolute and comparative vulnerability filtered through an assessment mechanism such as that provided by the IPCC. Beyond that, other criteria are important. Small island states and the least developed countries are already staking their claims based on the risk of sea level rise in the former and poverty and lack of development in the latter. Combining adaptation grants or donations with development loans would help ensure that adaptation actions are considered in the context of national development and strengthen capacity. GEF and its implementing agencies will also want to ensure that project proposals meet a high standard of preparation, and will obviously be influenced by evidence for win-win strategies, where climate change impact reduction benefits are additional to "what is worth doing anyway".

10. Some Conclusions

- There is no reason to challenge the *assumption* that considerable adaptation is possible at present levels of science, technology and wealth. There are reasons to fear that adaptation will fall short of its potential to reduce the impacts of climate change and variability. These include, at the international level, the lack of leadership, the absence of a common intellectual and policy framework, overt and sometimes destructive competition among international agencies, and deficiencies in the international system of governance or management.
- There is a complex equity issue both between developed and developing countries, and among the developing countries themselves. This can be expressed within the climate domain as a question of the prioritization of adaptation needs within developing countries and as the problem of who is particularly vulnerable between countries. This is set into the larger global issue of inequality between rich and poor countries, and the broader questions of trade and access to markets, debt and debt forgiveness and the rest.
- Even in the developed countries, the creation of effective policies for the mitigation of the impacts of climate variability and extreme events is a complex and difficult task that has failed to reduce losses. Reasons for this failure include a propensity for risk taking at the local and individual level propelled by market forces, the search for profit and amenity unhampered by weak public regulations and the ability of those who take the decisions that expose people and property to risk to avoid carrying the responsibility. None of this has so far been a sufficiently dramatic failure to generate the scale and type of actions needed at the national level. Despite the grief and anguish (noneconomic losses) inflicted on the victims of natural disasters, there is not yet an embodied culture of hazard and disaster avoidance. Perhaps the level of hazard and disaster losses is acceptable at the present level since the economic costs can be readily absorbed within

the national economy. In short, in the country with the greatest capacity to adapt and to reduce hazard and disaster losses by mitigation (the United States), the capacity has not been used to anything like its full potential.

- In many developing countries it seems probable that the impacts of current climate variability and extremes are an increasing proportion of national wealth, and it is clear that, in some well-reported instances, disasters have set back the process of development by a decade or more. The case for international assistance to help strengthen adaptive capacity is strong, and the list of needed policies and measures is long. Given the other pressing problems of development the issue of prioritization is very important. This applies to the selection of priorities within the climate impacts and vulnerability domain, and between the climate risks and other areas of public policy and investment. Hence the need to incorporate climate change adaptation policy within the framework of national economic development and physical planning. The science of adaptation in the different sectors (agriculture, managed forests, water resources, infrastructure, health, coastal zones, etc.) is well developed, so that an agenda for capacity building is not difficult to construct. On the basis of experience in the developed countries, especially the United States, there remains the nagging question, "If the adaptive capacity is created will it be used?" This is not an argument for doing nothing, but it is a caution that should influence the design of adaptation capacity building programs.
- The developed country parties to the Framework Convention are committed to the provision of assistance to help particularly vulnerable countries to adapt. They have legitimate concerns that adaptation funds should be spent efficiently and in a targeted way. The El Dorado of adaptation funds should not be used directly or indirectly for unintended purposes. The donor countries are also concerned about burden sharing, each country being reluctant to provide more aid than it considers to be its share. This applies especially to funds that are pooled through multilateral organizations. Donors also tend to have their own priorities for the direction of funds, either for historical or foreign policy reasons and may prefer to deal directly with their "client" states. At the same time the recipient countries are understandably sensitive to actual or perceived

infringements on their national sovereignty, and are resistant to conditions that would require them to undertake unpleasant reforms.

- The issues of normal climate versus anthropogenic climate change, and of global versus nonglobal benefits, remain in contention.
- These circumstances are creating a miscellany of *ad hoc* arrangements, which seem likely to continue and grow in the absence of a more organized international effort, guided by some common understandings, in concept and in policy.
- The present pattern of "muddling through" does not meet the needs of either developed or developing countries, or those of the international and donor organizations. Some new architecture is required that will establish a clearer, fairer and more transparent process, without, of course, adding to the burden of international bureaucracy.

11. A Suggestion

The foregoing conclusions are a preliminary diagnosis of the present state of health of the capacity to develop adaptation capacity at national and international levels. I hope that the paper sufficiently demonstrates the need for a more thorough exercise along similar lines that should be carried out by the principal stakeholders together with some independent voices to help ensure that the more difficult issues are not avoided.

The aim of such a study would be to produce proposals for a new and improved architecture for adaptation capacity building. This would clearly build on the undoubted strengths and capacities that already exist, and would promote adaptation capacity building in a more coherent and coordinated way.

More is involved than just the architecture. The purpose is to create a mechanism that would review adaptation capacity building and adaptation policy proposals in recipient countries and help to marshal funds for their implementation. For example, the least developed countries are now starting to prepare NAPAs, and these should be subject to review and endorsement by an appropriate body. The task exceeds the grasp of any single existing organization, and cannot be effectively carried out without the cooperation of many agencies, national, international and even local.

There are existing models for such an approach, composed and structured according to the nature of the problem. They include:

- the Global Alliance for Vaccines and Immunizations (GAVI)
- the Digital Opportunity Task Force of the G-8 countries
- the Heavily Indebted Poor Countries (HIPC) Debt Initiative
- the proposed Global Alliance for Basic Education (GABE).

If it seems utopian to speak of a global adaptation policy (GAP), then should we not at least go to work on a global adaptation framework (GAF)? The need is to formulate and advance generally acceptable guiding principles that can simultaneously serve as directives and also become benchmarks for measuring progressive accomplishments.

References

Abramovitz, J. N. (2001), Averting natural disasters. Chapter 7 in *State of the World 2001,* A Worldwatch Institute Report, Brown, L. R. et al., eds. W. W. Norton, New York, 123-142.

Burton, I. and Lim, B., eds. (2001), An Adaptation Policy Framework: Capacity Building for Stage II Adaptation. United Nations Development Programme Global Environment Facility, New York.

Burton, I. and van Aalst, M. (1999), Come Hell or High Water: Integrating Climate Change Vulnerability and Adaptation into Bank Work. World Bank Environment Department Papers, Paper No. 72. World Bank, Washington, DC.

Burton, I., Huq, S., Lim, B., Pilifosova, O. and Schipper, E. L. (2002), From impacts assessment to adaptation priorities: the shaping of adaptation policy. *Climate Policy* **2-3**, 145-159.

Goklany, I. M. (1999), Richer is more resilient: dealing with climate change and more urgent environmental problems. In *Earth Report 2000: The True State of the Planet Revisited*, Bailey, R., ed. McGraw Hill, New York, 155-187.

Goklany, I. M. (2000), Potential consequences of increasing atmospheric CO_2 concentration compared to other environmental problems. *Technology* **7S**, 189-213.

Homer-Dixon, T. (2000), *The Ingenuity Gap*. Alfred A. Knopf, New York.

Mileti, D. (1999), *Disasters by Design*. Joseph Henry Press, Washington, DC.

Smit, B. and Pilifosova, O. (2001), Adaptation to climate change in the context of sustainable development and equity. Chapter 18 in *Report of Working Group II. Climate Change 2001: Impacts, Adaptation, and Vulnerability,* McCarthy, J. J., Canziani, O. F., Leary, N. A., Dokken, D. J. and White, K. S., eds. Cambridge University Press, Cambridge, UK.

White, G. F., Kates, R. W. and Burton, I. (2001), Knowing better and losing even more: the use of knowledge in hazards management. *Environmental Hazards* **3**, 81-92.

8. The Legal Framework of Adaptation and Adaptive Capacity

Roda Verheyen
Research Unit Environmental Law
University of Hamburg, Germany

This paper provides an overview of the legal framework that influences adaptation choices and decision-making. The analysis of the provisions of the UN Framework Convention on Climate Change as well as national, regional and other international law finds that a substantive obligation on states to undertake adequate adaptation measures or to enhance adaptive capacity. The vulnerability of a system must be the starting point for assessing what adequate adaptation means and—from a legal perspective—countries have a wide margin of discretion when making adaptation choices. Cost-effectiveness is only one of many possible criteria by which such choices must be judged. At the same time, legal rules and principles could restrict the discretion of decision makers, for example, by demanding that impact assessments be carried out before implementing adaptation measures, or by ruling out certain measures altogether.

1. Introduction

The Third Assessment Report of the Intergovernmental Panel on Climate Change (IPCC; McCarthy et al., 2001) has established categorically that many vulnerable systems, both natural and human, will suffer significant or even irreversible damage due to climate change. Such damage will be caused both by large-scale, successive changes such as sea level rise and by extreme events such as more frequent and severe droughts and floods. Thus far, scientific studies about impacts, adaptation and vulnerability have focused on scenarios based on doubled CO_2 concentrations in the atmosphere, and have hardly investigated the impacts associated with the upper end of projected range of warming of 1.4-5.8°C by 2100 (Watson et al, 2001). The likelihood of warming at the higher end of the projections seems yet unresolved (see Forest et al., 2002; Knutti et al., 2002), but it is safe to say that climate change will have yet unknown adverse effects around the world. In many regions, climate change will forever change the way people live, their livelihoods and the natural systems that they rely on.

As a consequence, climate change also challenges legal and political concepts such as property rights, planning regulations and frameworks, natural resource management, state citizenship, governance structures, the role of private and public actors in protecting human beings and their infrastructure, and the principle of sustainable development. Thus, scientific research and discussion about adaptive capacity of developing countries and adaptation options for those countries and regions cannot take place in a legal (and socioeconomic) vacuum.

The climate change regime[1] within the UN Framework Convention on Climate Change (FCCC) of 1992 (UNEP/WMO, 1992) and the Kyoto Protocol of 1997 (UNFCCC, 1997) provides a framework in international law for the obligations of countries to implement adaptation measures and thus enhance adaptive capacity (Article 4.1 of the FCCC) and the financial obligations of developed countries to bear the costs of adaptation (Articles 4.3 and 4.4 of the FCCC). The latter issue is not discussed in this paper. Instead, a separate chapter has been dedicated to the issue of adaptation funding in this volume (Chapter 9). Yet, both themes must be considered together and, therefore, some cross referencing is necessary.

Apart from the climate regime, other international law principles and treaties as well as national and regional laws apply to climate change and its impacts and contribute to the broader framework in which adaptation measures can be undertaken or, as the case may be, must be undertaken (see Table 1).

In this context it is important to note that, while for the IPCC climate change consists of both natural variability and anthropogenic climate change, the provisions of the FCCC and Kyoto Protocol apply to only anthropogenic climate change as defined in Article 1.2 of the FCCC: "a change of climate which is attributed directly or indirectly to human activity that alters the composition of the global atmosphere and which is in addition to natural climate variability observed over comparable time periods". This definition does not, however, apply to other legal systems. In a national setting, the law will have to deal with climate change as defined by the IPCC and its impacts as a whole, and not only with anthropogenic climate change.

Table 1. The legal framework for adaptation decisions: encouraging adaptation or slowing/hindering the implementation of measures by legal rules applicable on different levels (international, regional, national)

Types of rules	International law	Regional law	National/ communal laws
Obligation to undertake adaptation measures (encouraging measures)	Human rights conventions, nature protection (Ramsar, CBD etc.), international climate regime	Various regional conventions (e.g., on watercourses, other natural resources)	Civil law (to protect property/water/land rights), administrative law (to protect, e.g., human health), etc.
Obligation to finance adaptation (encouraging)	Climate regime, other public international law	Regional human rights treaties	Constitutional law (obligation of a state to protect its citizens)
Liability/responsibility for residual damages (neutral)	Customary international law (but: enforcement problem)	Regional human rights treaties	Recovery via civil law, constitutional law, administrative law
Regulations restricting/prohibiting adaptation choices (hindering implementation of certain measures)	Few nature protection conventions, human rights (in case of resettlement), etc.	Treaties regulating the allocation of natural resources, nature conservation	Conflicting property rights, social/human rights, land rights, nature conservation laws
Regulations guiding adaptation choices (impact assessment, planning regulations) (slowing down, changing choices)	Customary law/ treaty obligation to conduct environmental impact assessment	Customary law/ regional treaties on environmental impact assessment?	National planning codes, impact assessment legislation, sustainability objectives, other policy objectives

This is the context in which this paper outlines the basic principles and legal obligations, and introduces questions and topics for further research. Compared to the issue of mitigation, research about adaptive capacity in developing countries, the linkages between development and vulnerability, and "best practice" adaptation measures have been of minor importance in the scientific and legal literature, international negotiations and national climate change activity. Also, the focus of the international negotiations has been to prevent dangerous anthropogenic climate change and thus to achieve the objective of the FCCC (Article 2). However, at least on the international level, the focus on mitigation is changing now that decisions on the implementation of the Kyoto Protocol's emission reduction commitments were made at the Seventh Conference of the Parties (COP7) at Marrakesh in November 2001.[2] The future research agenda in this area will thus also significantly influence the political and legal developments over the next decades.

But the state of development of the law on adaptation (i.e., infancy) is also because "adaptation" or "enhancing adaptive capacity" can be achieved in so many different ways in different regional and national settings, it has been difficult to provide an abstract legal framework for the necessary decisions. Some brief comments about the terminology, which is also used in Chapter 9 on adaptation funding, are in order.

1.1 Terminology

For the context of this paper, the term adaptation is used to describe the adjustment of ecological and socioeconomic systems—as a result of processes, measures and structural changes—to current or projected climate changes. Furthermore, adaptive capacity is defined as the ability of a system to adapt to climate change, to moderate potential damages, to take advantage of opportunities or to cope with the consequences (Box SPM 1 in McCarthy et al., 2001). The adaptive capacity of a system depends on its vulnerability, which is a function of exposure to climate change hazards and the capacity of the system to adapt. Reducing vulnerability thus enhances the adaptive capacity of a system and vice versa (see Chapter 2 of this volume).

There are various ways of adapting to climate change and I will not repeat what is discussed elsewhere in this volume. It suffices to say that the distinction between anticipatory and reactive adaptation of natural or human systems (see Smit et al., 2000; McCarthy et al., 2001; Klein, in press) is useful for the legal analysis, and so is the distinction between autonomous and planned adaptation. Autonomous adaptation will not involve decisions that can be tested against applicable law beforehand, but will be triggered by market or natural changes. Planned adaptation involves legally relevant decision-making.

In this analysis (as well as in Chapter 9 of this volume) I also distinguish between measures to enhance adaptive capacity and adaptation measures, by thinking in terms of macro and micro measures. To enable a system to prevent climate change damages more effectively, improved housing and infrastructure, education, dissemination of information and generally the eradication of poverty could prove very successful; such measures take place on a macro level. Micro adaptation measures include increasing the height of sea walls, building irrigation systems to protect agriculture from salt water intrusion, or installing air-conditioning systems to avoid heat-related illness or death.

1.2 The Legal Framework

There are several ways of describing the legal framework in which adaptation decisions will take place. On one side of the coin are obligations of countries to adapt to climate change to prevent damage to natural systems or human beings or to finance adaptation measures undertaken by others. It is this legal framework the international climate regime is concerned with. In addition, other international law rules such as human rights, nature protection and national (constitutional) requirements provide thresholds and detailed rules about when and to what extent adaptation measures might be necessary or mandatory to protect human beings, the environment, natural resources, etc. In this context, synergies between different international regimes have a role to play, for example, between the climate regime and the UN Convention to Combat Desertification (CCD) and the UN Convention on Biological Diversity (CBD), both of 1992.

On the other side of the coin are legal rules prohibiting or restricting adaptation choices. In the implementation process of adaptation measures or measures to enhance adaptive capacity, decision makers have to take into account the various legal frameworks on national, regional and international levels. While it is difficult to predict the legal conflicts that might arise, planned adaptation will generally be subject to legal and political processes involving planning multiple uses of natural resources (such as land and water) and preventing conflicts. It is also conceivable that strict rules prohibiting certain measures and international or national rules on environmental impact assessment could apply to adaptation measures. For example, the FCCC refers to the need for environmental impact assessments in its Article 4.1(f), but does so in a nonbinding manner. Binding rules could be contained in international, regional or national legal instruments (for example, nature protection requirements if a certain ecosystem would be destroyed to protect human infrastructure, or human rights instruments (in cases where a measure would, for instance, entail the displacement of peoples).

While adaptation can significantly reduce vulnerability to climate change impacts, no level of adaptation will avoid impacts and residual damage from climate change completely. Thus, the issues of responsibility and liability for damages caused by climate change add another legal dimension.

Admittedly, this paper is naturally biased toward the framework provided by international law, and in particular the international climate regime. In terms of domestic laws that would guide adaptation decisions in developing countries, I can only sketch some contentious issues and hope that the ongoing activities preparing for adaptation decisions in the future—national adaptation programs of action (NAPAs), vulnerability assessments, national communications—will take into account, to an appropriate extent, the existing legal framework in the country in question. Naturally, a legal analysis cannot address socioeconomic and political issues associated with adaptation choices that must be addressed in practice by decision makers. I highlight and discuss legal issues only, which can hopefully assist successful decision-making. International law, while difficult to enforce, in practice still binds decision makers and, particularly in the field of climate change, provides a direction to national and regional law makers. For this purpose, it is an important tool that should be used.

2. Obligation to Undertake Adaptation Measures or Enhance Adaptive Capacity

2.1 The International Climate Change Regime

While both the FCCC and the Kyoto Protocol focus on the prevention of dangerous climate change, they also set a framework for adaptation activities. (For a detailed overview of the FCCC, see Sands, 1992; Bodansky, 1993; Sprinz and Luterbacher, 1997; and Verheyen, 1997. For an analysis of the Kyoto Protocol and its history, see Grubb et al., 1999, and Oberthür and Ott, 1999.)

Obviously, mitigation and adaptation are two sides of one coin, complementing and depending on each other. Should the international community be able to achieve the ultimate objective of the FCCC, which is "stabilization of greenhouse gas concentrations in the atmosphere at a level that would prevent dangerous anthropogenic interference with the climate system" (Article 2), adaptation would not be necessary. However, it is unlikely that, as the FCCC stipulates, stabilization of greenhouse gas concentrations at a sustainable level will be "achieved within a time-frame sufficient to allow ecosystems to adapt naturally to climate change . . ." (Article 2).

The vulnerability of developing countries in particular was recognized explicitly before and during the negotiations of the Convention (see UNGA Res. 44/206, 22 December 1989, UN Doc. A/44/862, para. 4 (available at www.un.org). Some even called for specific legal instruments to tackle the issue of adaptation needs. Germany, for example, proposed a protocol on "the adjustment to climate changes and the prevention and containment of climate-related damage" (UN Doc. A/AC.237/Misc.1/Add.1, "Set of informal papers provided by delegations, related to the preparation of a framework convention on climate change" at 21). These early attempts, however, were reduced to rather ambiguous treaty language in Article 4 of the FCCC, which also contains the "soft" emission reduction obligation for industrialized (Annex I) countries.

2.1.1 The Core Rule: Article 4.1(b) of the FCCC

Article 4.1(b) of the FCCC provides that all Parties (including developing countries) must "formulate and implement national or regional programmes containing measures to mitigate climate change . . . and measures to facilitate adequate adaptation to climate change". The latter part of this provision has been used by Parties mainly in conjunction with the duty obliging all countries to report about their efforts to implement the Convention (Article 12.1 of the FCCC), which applies to developing and industrialized countries alike. It complements Article 3.3 of the FCCC, which calls on Parties to "take precautionary measures to anticipate, prevent or minimize the causes of climate change and mitigate its adverse effects".

Furthermore, Article 4.1(e) calls for the preparation for adaptation to the impacts of climate change. To this end Parties shall "cooperate in preparing for adaptation to the impacts of climate change; develop and elaborate . . . plans for coastal zone management, water resources and agriculture, and for the protection and rehabilitation of areas, particularly in Africa, affected by drought and desertification, as well as floods". It is notable that the Convention draws particular attention to Africa, especially since FCCC-related adaptation activities have been relatively scarce in this region.

As programs are implemented by carrying out specific, real-life measures to facilitate adequate adaptation, the Convention clearly contains a legal obligation for Parties to address adaptation in a strategic way through programs and, importantly, also obliges them to implement them. Two questions arise here: what exactly is meant by "facilitate measures", and what constitutes "adequate adaptation"? These questions will certainly become more and more contentious as adaptation requirements and lack of adaptive capacity in developing countries are subject to further in-depth research, and will require some practical solutions. Also, the provision theoretically applies only to adaptation to anthropogenic climate change, since the FCCC's scope is restricted to human activities. This is an issue I return to below and which is also discussed in Chapter 9 on adaptation funding.

Facilitation of Measures. Within the climate regime, the term facilitation seems to mean something other than the full catalogue of measures that could be envisaged to make a system climate-safe. Literally, facilitate means "to help, aid or to assist", and in practice it is often used in conjunction with planning or "enabling activities" for categorizing funding provided by the Global Environment Facility (GEF). However, based on the Convention text, there is no reason to conclude that the programs mentioned must be confined to any particular type of measures. Rather, they could include the full range of measures that would enable "adequate adaptation". Since adaptation is understood to be a process rather than any particular measure,[3] methods discussed by the scientific community to enhance adaptive capacity in developing countries, such as macro-level instruments in the health, education or housing sectors, should qualify as measures facilitating adaptation.

Adequate Adaptation. The FCCC fails to define the term "adequate". It appears in Article 4.1(b) and again in Article 4.3 referring to the adequacy of the flow of funds. Under Article 4.1(b), Parties are not obliged to carry out any particular kind or type of measure. The article does not specify the time period in which the duty arises, nor does it set a deadline for either the formulation of the adaptation programs or their implementation. However, measures that infringe on other international or national rules could be deemed inadequate at the outset. Also, adaptation measures that would result in long-term greenhouse gas emissions can be deemed inadequate in the context of the Convention because they would operate contrary to Article 2.

Logically, the process to determine what is adequate must start with a vulnerability assessment of the pertinent region or community. Criteria for the determination of the adequacy of measures could include economic efficiency, environmental sustainability, technical feasibility, administrative/legal admissibility and social and political acceptability of the measures (Klein and Tol, Adaptation to Climate Change: Options and Technologies, Technical Paper FCCC/TP/1997/3, 29; available at www.unfccc.int).

At various occasions during the international climate change negotiations, developed states (donor countries) have suggested that adaptation measures are adequate only if they are cost-effective, for instance, when the system at risk is more valuable than the overall cost of the considered measures. However, determining cost-effectiveness in this sense is riddled with uncertainties, knowledge gaps and elements of judgment (Klein, in press). The economic toolbox alone will not provide the answer to problems of judgment when it comes to enhancing adaptive capacity and to adaptation measures themselves. Rather, the choice of adaptation options will depend very much on the special circumstances of a country or region. Legally, Article 4.1(b) of the FCCC gives countries formulating and implementing adaptation programs discretion to judge what is adequate. There is no provision in the FCCC to prohibit a country from making a choice that proves to be not cost-effective when applying economic tools, for example, when prioritizing the protection of a particular ecosystem over the preservation of infrastructure. Such (noneconomic) decisions could in fact be mandated by international or national law. Authors who have argued against the sole use of economic tools such as cost-benefit analysis to determine adaptation options (Callaway et al., 1998; Klein, 2001) thus find support in international law. This should be taken into account when interpreting the financial requirements of the FCCC and when delivering funding for adaptation purposes.

The FCCC does not oblige Parties to take every possible measure to prevent climate change damages. Rather, by using the undefined term "adequate", the Convention leaves a margin of discretion for each Party to choose between preventing residual damages and accepting them. Whether or not this choice entitles any country to compensation cannot be explored here, but it is conceivable.

For the benefit of developing countries, future negotiations and research should provide some answers to the questions raised above to ensure the early integration of these issues into general development planning and to enable full implementation of the financial obligations in Articles 4.3 and 4.4 of the FCCC (see Chapter 9 of this volume).

Private or Public Adaptation? A mostly unexplored but important issue is the role of the private sector in the framework of enhancing adaptive capacity and implementing adaptation projects. Since the Convention as an international treaty cannot address private entities directly, Article 4.1(b) of the FCCC governs only public adaptation measures and does not prescribe any particular activities by private entities.

It has been argued that "social capital" is indeed just as important for the adaptive capacity of a system as technical means or formal governance structures. Social capital is generally defined as features of social organizations such as trusts, norms and networks that can improve the efficiency of society by facilitating coordinated actions (Chapter 3 of this volume). The capacity of parts of society to act collectively in the interest of all is particularly important in the absence of effective governance structures. This proposition begs the question to what extent the climate regime encourages (private) informal networks and communities to act collectively to enhance adaptive capacity.

While Article 6 of the FCCC recognizes the role of civil society in addressing the impacts of climate change and explicitly calls on states to enable "public participation in addressing climate change and its effects and developing adequate responses", there is no explicit acknowledgement in the Convention of the role of the private sector in adapting to climate change. However, in many vulnerable economies, the private sector provides important industrial and general infrastructures and will feel responsible (or even be legally responsible) for any necessary protection. In fact, private companies might hold the means (technological, financial and human capacity) to actually increase the adaptive capacity of communities in developing countries.

Private and public adaptation measures will complement each other in practice. However, private entities cannot seek funding under the Convention for their engagement. Rather, this right is reserved for developing country Parties (see Chapter 9 in this volume), who may decide to include measures to support the private sector, local communities or households in their programs under Article 4.1(b) and thus also in applications for funding under the Convention's financial mechanism. In fact, the potential of the private sector should also play a role in determining whether an adaptation option is "adequate".

2.1.2 Implementation by Developing Countries—National Communications and NAPAs

Article 4.1(b) of the FCCC has been implemented with different degrees of effectiveness in developing countries. This is partially because the circumstances under which activities would be funded by the GEF as the financial mechanism of the Convention are unclear, but also because implementation is so inextricably linked with general economic and social development. Because of the real lack of country data as a basis for decision-making, adaptation planning in most developing countries is still in the assessment phase. As such, many countries currently are only compiling information for adaptation decision-making as part of the development of their national communications (pursuant to Article 12.1 of the FCCC), or are just entering the planning phase.

By the end of 2001, only 52 out of 146 non-Annex I had submitted their national communication, in which they also provided information about the impacts of climate change and adaptation (FCCC/SBI/2001/14/Add.1, available at www.unfccc.int). The methodological approaches used by Parties were generally consistent with the analytical framework provided in the IPCC Technical Guidelines (IPCC, 1994), which Parties are encouraged, while not obliged, to use (Decision 9/CP.2 (Annex 1 Parties), Annex, para. 41 and Decision 10/CP.2, (non-Annex 1), FCCC/CP/1996/15/Add.1., see www.unfccc.int).[4] The utility of these guidelines is questionable, in particular when it comes to linking national adaptation assessments to the legal obligations of countries, which includes the financial duties of Annex II countries under Articles 4.3 and 4.4 of the FCCC (see Chapter 9). The application of the guidelines seems to produce basic country assessments and requirements, but does not provide a methodology or legal tools for decision-making. Because demand for adaptation funding will grow, there is a clear need for guidelines that enable decision makers to prioritize activities or projects. The new guidelines adopted at COP8 do not seem, at first glance, to remedy this problem.

To ensure that adaptation science actually supports the enhancement of adaptive capacity, it is imperative that future research focuses on providing tools for decision-making within the international climate regime. In this context it is also important to ensure that planning and programming activities are linked with other, similar processes under other Conventions or general sustainability planning.

The lack of implementation of Article 4.1(b) as well as the unavailability of funds for adaptation purposes in particular for least developed countries (LDCs, who also lag behind most in developing their national communications, compelled Parties at COP7 to adopt the concept of NAPAs (Decision 27/CP.7 and 28/CP.7, contained in FCCC/CP/2001/13/Add.4).[5] NAPAs are designed to communicate the urgent adaptation needs of LDCs, and they are the only specific document within the framework of the climate regime to address exclusively adaptation.

The gist of the NAPAs will be a list of priority activities necessary for adapting to current and projected climate change impacts in the respective country. Activities are to be prioritized in the NAPAs if delays in implementation such projects or other activities[6] "could increase vulnerability, or lead to increased costs at a later stage" (FCCC/CP/2001/13/Add.4, p. 8).

Pursuant to the initial guidelines on NAPAs issued by COP7, the programs should be prepared in cooperation with local communities and other stakeholders, be multidisciplinary in character and build on other action plans under the CCD (available at www.unccd.int/convention/menu.php) and the CBD (available at www.biodiv.org/convention/articles.asp). A national NAPA team will be set up by the national climate change focal point to organize a consultative process leading to the preparation and implementation of the NAPA. As a first step, the guidelines call for a participatory assessment of vulnerability to current climate variability and extreme weather events as well as an assessment of potential increasing risks induced by climate change. On this basis, NAPA teams are to identify country-driven criteria for selecting priority activities and to then suggest priority activities using a set format. It is striking that the legal differentiation between natural and anthropogenic climate change seems to have been abandoned.

A preliminary list of criteria is contained in the guidance; they include level or degree of adverse effects of climate change, poverty reduction to enhance adaptive capacity, synergy with other multilateral environmental agreements and cost-effectiveness. They are to be applied to a set of societal values or functions, such as loss of life, human health, food security, water availability, etc. The final NAPA document should be endorsed by the government of the respective country since the list of activities also serves as the basis for seeking financing through the LDC fund, i.e., from the GEF.

It is questionable to what extent these criteria will help the national NAPA teams determine adaptation needs to be captured in a "list of priority adaptation activities". However, the guidelines recognize various needs: the need for integration of adaptation measures into other activities, the need for structural measures to increase adaptive capacity, and the fact that current vulnerability can give important clues to future vulnerabilities. Moreover, with the NAPA guidelines adopted at COP7, Parties have recognized that the reduction of current vulnerability to climate variability can lead to better adaptive capacity in LDCs in the decades to come. Therefore, it is unlikely that response activities will be strictly limited to impacts caused by anthropogenic climate change. This is an important deviation from the principles in the funding provisions of the FCCC and might result in a situation in which the climate regime provides funding for measures that are beyond its initial scope. NAPAs are also a first step toward using vulnerability assessments, which focus on the system of interest as the basis for decision-making, rather than impact assessments.

2.2 Other Legal Instruments

Several other legal instruments oblige states, private entities or individuals to take preventive action against the impacts of climate change. First in line are human rights instruments, such as the 1948 Universal Declaration on Human Rights (UNGA Res. 217 A (III), available at www.un.org/law), the 1966 Covenant on Civil and Political Rights (6 ILM (1967) 368), the 1969 American Convention on Human Rights (9 ILM (1970) 673 and 36 ILM (1997) 229) and the 1981 African Charter on Human and Peoples Rights (213 UNTS 221).[7] These are mainly aimed at securing human rights of citizens within the borders of their home country. Thus, they might oblige states to take preventive action to ensure that human rights such as the right to life, property rights, the right to health etc., are not infringed as a result of climate change, be it anthropogenic or natural. Evidently, these kinds of obligations will find their limit in what a state can reasonably be expected to deliver to its citizens in terms of finance and protection. Developing nations, even if they are parties to human rights treaties, are not likely to find answers to difficult adaptation choices in the letters of those legal texts. This is also clear from experience with natural disasters, floods and storms in the past—developing countries are often not able to prevent significant losses, and legal obligations have rarely played a constructive role in triggering effective measures.

Against this background, some have argued that foreign states (i.e., states that might be able to afford certain protection measures) should be bound by human rights instruments when they somehow affect citizens of another country, as would be the case with man-made climate change (see Gibney et al., 1999). The authoritative Inter-American Commission on Human Rights heads in the same direction when it argues that ". . . given that individual rights inhere simply by virtue of a person's humanity, each American State is obliged to uphold the protected rights of any person subject to its jurisdiction. While this most commonly refers to persons within a state's territory, it may, under given circumstances, refer to conduct with an extraterritorial locus where the person concerned is present in the territory of one state, but subject to the control of another, usually through the acts of the latter's agents abroad . . ." (Coard v. United States, Report No. 109/99, Case 19.951, 29 September 1999, available at www.cidh.org). If this opinion were to prevail, a country might be obliged to conduct measures to, for example, protect a specific community against more severe and frequent weather extremes that are due to anthropogenic climate change. Such a case would raise issues of causation and responsibility similar to the ones discussed below.

It should be noted though, that human rights principles are based on the presumption that a state owes certain obligations to individual or specific groups of citizens. To fulfill this obligation, a state can be required to either act or omit a certain activity—the measure must be relatively concrete. It is unlikely that human rights rules would capture measures that are generally aimed at enhancing adaptive capacity of developing countries (so-called macro measures) rather than having a specific protective purpose (so-called micro adaptation projects). This dynamic area of law should be explored further as the impacts of climate change become more and more apparent.

Similarly, many national constitutions contain human rights provisions that are very well developed and enforceable in some countries, such as in India. Should developing nations chose to either ignore adaptation to the impacts of climate change as part of their official development, infrastructure and planning activities, or adopt only adaptation measures that benefit specific parts of the population or industry, it might be possible for citizens to challenge such behavior in the courts of these countries. This is also a situation where specialized international and national rules to protect indigenous peoples might apply. Given the reliance of indigenous peoples on natural resources, they are particularly vulnerable to both incremental climate change and extreme weather events.

3. Legal Principles and Rules Guiding, Prohibiting or Impeding Adaptation Activities

The objective of "adaptation to climate change" as contained in the FCCC and Kyoto Protocol can be achieved only if synergies are explored and exploited. This goes two ways: reducing vulnerability and enhancing adaptive capacity calls for sustainable development practices and thus a redefinition of development planning and the scope and purpose of development aid. At the same time, measures that might contribute to making a system climate-safe in the face of anthropogenic as well as natural climate change will impact on areas that are regulated by environmental law themselves, such as biological diversity, international watercourses, desertification, etc.

Looking at a list of possible measures to enhance adaptive capacity or discrete adaptation projects, such as new building codes, restricted planning in threatened areas, relocation, raising of sea defenses, construction of house on stilts, irrigation systems, tree planting and land leveling, it is clear that some measures might impact substantially on natural and human systems. This is acknowledged implicitly by, for example, Article 4.1(f) of the FCCC.[8] Obviously, some measures might also support objectives of other legal instruments such as biodiversity or water conservation.

It should be noted at the outset that international law will seldom provide decision makers with hard rules that have to be applied. Such rules will more often be found in national and perhaps regional laws, which are often less developed and strict in developing countries. This does not mean that there are no strict environmental laws in developing countries. However, the density of regulation of activities varies and, in LDCs in particular, the legal structure for decision-making is far less developed than in industrialized countries.

I would like to emphasize that some measures to enhance adaptive capacity of a developing country might contradict the primary purpose of the climate regime itself—to prevent dangerous climate change. Projects that result in long-term greenhouse gas emissions would thus—in my view—be per se in conflict with international law and therefore also inadequate in the sense of Article 4.1b of the FCCC.

3.1 International Regimes—Conflicts and Synergies

On the international level, environmental regimes have been shown to generally reinforce and support rules and activities upheld by another regime, whereas others display relationships of normative conflict, counteracting efforts or duplication of activities (Stokke, 2001). And while scientific interest in interregime relationships is fairly recent, there is broad agreement about the vast scope of better management of the cross-institutional relationship to enhance the effectiveness of those regimes. For the purposes of this paper, some nonclimate related regimes are pertinent, in particular the CBD and CDD of 1992, but also various regional watercourse treaties such as the Ramsar Convention on Wetlands of 1971 (available at www.ramsar.org).

The most obvious and best studied regime interplay is the one between the FCCC and CBD, mostly because insufficient action to halt climate change will have severe impacts on biodiversity and because of the role of the terrestrial biosphere in storing carbon. This has been noted by the 5th Conference of the Parties to the CBD (Decisions V/3, V/4, V/15 and V/21, available at www.biodiv.org). The CBD subsidiary body for scientific and technical advice established a technical expert group whose task it is to prepare scientific advice to integrate biodiversity considerations into the implementation of the FCCC and its Kyoto Protocol by undertaking a pilot assessment in cooperation with the FCCC bodies. With regard to adaptation in particular, both negative and positive interplay can be envisaged here: the IPCC Working Group II (McCarthy et al., 2001) has suggested some adaptation options that actively contribute to the conservation and sustainable use of biological diversity. These include expanding aquaculture to relieve stress on natural fisheries; designing terrestrial and marine multiple-use reserves and protected area that incorporate corridors which would allow for migration of organisms as a response to climate change; efficient use of natural resources; small-scale restoration of inland wetlands; restoration of degraded soils especially in rangelands; and adjustments in the timing and intensity of livestock grazing.

It is probably fair to say that—at least when seeking biodiversity conservation—adaptation activities that conserve or restore natural ecosystems are generally positive, while activities that use hard technologies and that change natural ecosystems are generally negative. For example, defending a country's coast against sea level rise can save ecosystems, but also have very negative impacts on coastal species and get in the way of the "natural development" of such regions. Similarly, building defenses against extreme weather events could be achieved by replacing vulnerable ecosystems with other, artificial systems (flood protection, storm shelters, etc.) (Öko-Institut, 2001). However, since the CBD has not further defined what type of biodiversity it aims to protect (ecosystem or species or genetic diversity?), legal rules to be followed by decision makers are almost absent. There is, for example, a lack of criteria for sustainable forest management, so if enhancing adaptive capacity would touch on forest ecosystems, there are no "hard standards" against which such measures could be tested. However, one could also conclude that such standards are not needed for many measures leading to better adaptive capacity in developing countries, since those can be (on their face) neutral with regard to nature protection. This is particularly true for macro measures such as better education and eradication of poverty.

The Ramsar Convention on Wetlands of 1971 (available at www.ramsar.org) seeks to ensure "the conservation and wise use of wetlands by national action and international co-operation as a means to achieving sustainable development throughout the world". The FCCC and Ramsar Convention are involved in an ongoing process to identify the role of wetlands for the climate system (Ramsar Resolution VII.4,7, Resolution VII.4,13 and Recommendation 7.1.5, available at www.ramsar.org; and FCCC/SBSTA/1999/14, available at www.unfccc.int). The adaptive capacity of wetlands has not been explored in depth, which could provide a trigger to incorporate the Ramsar list of wetlands with global importance into the climate regime (see Bergkamp and Orlando, 1999).

There is potential for synergies as well as conflicts between both conventions: for flood protection, rehabilitation and restoration of natural wetlands are recommended, which would provide mutual benefits under both Conventions. Potential conflicts between objectives of the Ramsar Convention and objectives under the FCCC could arise from activities to adapt to sea level rise in coastal zones, especially from so-called hard adaptation technologies such as the construction of dikes, levees and floodwalls, floodgates and tidal barriers, the prevention of saltwater intrusion in coastal estuaries and groundwater aquifers using barriers, and the artificial movement of freshwater inlets further upstream.

With regard to regional water treaties, of which there are hundreds, seeking to regulate the use, protection and allocation of fresh- and groundwater (see for an overview Sands, 1995, and the 1997 UN Convention on the non-navigational use of watercourses, 36 ILM (1997) 700, which seeks to provide an overarching framework to these regional and bilateral treaties), enhancing adaptive capacity will often support the aims of those treaties since they are concerned with the management of the natural resource the streams provide. However, should a country build irrigation systems to meet a region's increased water demands due to climate change, this could conflict with water allocation as agreed in regional treaties. In such cases, a country might intend to adapt for the benefit of its population and in the result, create a resource conflict with another riparian use.

Generally, it seems that research on the international law level should focus on the synergies between and mutual benefits of regimes rather than on potential conflicts (see Miles et al., 2001). This does not mean, however, that the opaqueness of terms and concepts of one regime should simply be shifted to another. For example, Klein (2001) suggests an "ecosystem approach to adaptation", although it is unclear what this approach means in practice and in what way it could provide a means to balance different objectives (adaptation/conservation) in the framework of a concrete project or planning proposal. The ecosystem approach was adopted by the Parties to the CBD as the primary framework for action under the Convention (Decision II/8 CBD and Decision V/6 CBD), but implementation has been very slow.

Another typical—more procedural—question arises when projects are planned or undertaken: should adaptation measures and measures to enhance adaptive capacity generally be subject to an environmental impact assessment (EIA) or to a broad sustainability screening? Given the possible impacts of adaptation projects based on technology, this should certainly be considered. EIAs are mentioned, but not made mandatory in Article 4.1(f) of the FCCC and encouraged by Article 14 of the CBD. Article 14 of the CBD requests EIAs of proposed projects in order to avoid or minimize adverse effects on biological diversity and to allow public participation in such procedures. Similar provisions might be adopted in the framework of the Kyoto Protocol in order to avoid negative impacts on biodiversity from land-use change and forestry projects.

It is questionable whether EIAs are part of customary international law. Rather, they seem to be an element of regional international law, as codified by the 1991 UN ECE Espoo-Convention (30 ILM (1991) 802), which establishes mandatory EIAs for projects with transboundary impacts. Adaptation projects might have such transboundary impacts and thus international law could apply, but they might affect only the country in question. Article 14 of the CBD thus seeks to expand the definition of transboundary effects: in the rationale of this Convention, all impacts on species or genetic diversity are a matter of common concern. In this respect, the law remains opaque.

However, because many adaptation policies and projects will involve international finances, provided by industrialized states and channeled through international institutions, therc might be a legal obligation on donor countries to scrutinize the impacts of such funding. Such obligations are still being discussed by academics, while some institutions such as the World Bank have already adopted their own guidelines on mandatory impact assessment. It seems clear that in the absence of clear legal rules, guidelines to this effect should be adopted within, for example, the climate regime.

3.2 Regional and Domestic Law

Both developing countries and industrialized countries possess relatively detailed environmental laws protecting biodiversity, water and soil quality. At the same time, property and land law will often interfere with adaptation choices. Essentially the same rules that could require adaptation measures for the benefit of human rights or indigenous peoples could also make certain choices illegal. However, there is little point in hypothesizing about potential conflicts in the abstract. Rather, such conflicts can be prevented either by assessing the legal framework along with the vulnerability of a certain system before determining a measure to be taken or by trusting that regional or local government in the respective country will ensure that legal obstacles to the implementation of projects or measures will be removed. The latter option is obviously not always ideal, but the reality in sovereign nations in a long-term process of economic and development planning.

4. Responsibility/Liability

As recent assessments and models show, adaptation measures and measures to increase adaptive capacity are capable of significantly reducing damages to human infrastructure and natural systems alike. However, it is also clear that projected climate change will result in adverse impacts that cannot be prevented, in particular in developing countries and lower latitudes generally (residual damages).[9] This raises the issue of who is legally responsible for such damages.

In this context, adaptation measures and measures to enhance adaptive capacity can be defined as direct damage prevention because they can reduce residual damages.[10] Legally, the obligation to prevent damages directly corresponds with the obligation to compensate for any damages done.[11] Since humankind is committed to some climate change regardless of mitigation efforts, both are possible responses to a situation in which the rights or interests of states or/and individuals are affected because of ongoing and past activities (in this case, emissions of greenhouse gases). Therefore, if general legal responsibility for climate change damages is established, such obligation also covers adaptation measures (and costs) as direct damage prevention measures over and above the financial obligations of the international climate regime.

One of the basic rules of international law is that states shall not inflict damage on other states. In environmental law, this rule is captured in the so-called "no harm principle", which in turn has its foundations in the principle of good neighborliness between states, which are formally equal under international law. (See the famous Trail Smelter Arbitration of 1941, Reports of International Arbitral Awards (RIAA) III (1941) p. 1905, and for further analysis, Perrez, 1996.)

Principle 2 of the 1992 Rio Declaration (available at www.unep.org) reiterates this rule of customary international law,[12] prohibiting transboundary environmental injury: "States have, the sovereign right to exploit their own resources pursuant to their own environmental and developmental policies, and the *responsibility* to ensure that activities within their jurisdiction or control *do not cause damage* to the environment of other states or of areas beyond the limits of national jurisdiction" (emphasis added).

Moreover, in international law, states are responsible for violations of public international law and are obliged to compensate the indirectly or directly affected states for the damage caused (see the landmark case Chorzów Factory, Permanent Court of International Law (PCIL) Reports Series A Nr.7, (1927), p. 30, and Tomuschat, 1999). This rule forms the basis of the law of state responsibility, a body of law that has recently been codified by the International Law Commission (ILC), a UN body entrusted with promoting the codification and development of international law. The ILC' s Draft Articles on State Responsibility (UN Doc. A/56/10, Report of the International Law Commission, 53rd session, available at un.org.law/ilc/) are the first comprehensive codification of the law on state responsibility. While these do not automatically represent international law but have to be accepted (e.g., ratified) by states, they can serve as a useful tool to examine the conditions and consequences of state responsibility for climate change damages. Under these rules, only states could challenge other states, or states could act on behalf of their citizens. (See the Report of the International Law Commission, 53rd session and www.un.org/law/ilc/archives/statfra.htm; and for a commentary, Crawford et al., 2001.)

In and before the negotiation phase of the FCCC, legal and political experts requested that "appropriate principles of liability, compensation or other relief" should be developed and that a "World Atmosphere Trust Fund" be established (Protection of the Atmosphere: Statement of the Meeting of Legal and Policy Experts, Ottawa, 22 February 1989). This fund would have helped developing countries to, inter alia, mitigate climate damages. However, both the IPCC topic group on Legal Measures, whose report served as a basis for early drafts of the FCCC, and the Convention itself avoid dealing with the topic of liability and responsibility. This stands in contrast to the relatively detailed provisions on adaptation to climate change and respective financial obligations. This fact caused several states, upon signature of the FCCC, to make the following declaration, which refers to state responsibility: ". . . signature of the Convention shall in no way constitute a renunciation of any rights under international law concerning state responsibility for the adverse effects of climate change . . ." (declarations made by the governments of Nauru, Tuvalu, Fiji and Papua New Guinea; see UN Doc. COP1.Inf02, Status of Ratification of the UN FCCC (24/3/1995) available at www.unfccc.int). This reservation had been proposed by the Alliance of Small Island States for inclusion in the Convention itself during the negotiations but was not included in the final document. The Kyoto Protocol does not contain any pertinent provisions either.

Thus, in the absence of any treaty law on the issue of residual damages, a claim under international law involves at a minimum the following steps: identifying the damaging activity attributable to a state, establishing a causal link between the activity and the damage, determining either a violation of international law or a violation of a duty of care (due diligence), which is owed to the damaged state, and quantifying the damages caused by the activity.

There have also been attempts to apply the "polluter pays" principle, originally an economic allocation concept, between states. Thus responsibility for environmental damages would be generated by the mere causation of damage—without proving any failure to act with due care. Interestingly, during the negotiations of the FCCC, it was suggested that the principle could serve as an appropriate legal framework to address issues of liability and compensation (Report of the INC, 1st session 4-14 February 1991, UN Doc. A/AC.237/6, 6 f. and UN Doc. A/AC.237/Misc.1/Add.3,24; available at www.unfccc.int). But so far, the legal content of the polluter pays principle remains unclear and it is doubtful whether it can be regarded as customary international law rule.

If one presumes that following IPCC statements are true (Watson et al., 2001) and that, according to some models, adverse impacts of climate change might range up to 5 percent of GDP in some developing countries, assuming only the doubling of CO_2 in the atmosphere (Tol and Verheyen, 2001), the facts seem to match the described legal framework:

- Most of the observed warming over last 50 years is likely (67-90 percent chance) due to increases in greenhouse gas concentrations due to human activities.
- About three quarters of the anthropogenic emissions of CO_2 to the atmosphere during the past 20 years is due to fossil fuel burning.
- From the collective evidence, there is high confidence (67-95 percent) that recent regional changes in temperature have had discernable impacts on many physical and biological systems.

This is especially true for damages in developing countries, which have—in the past—hardly contributed to the increase in atmospheric carbon dioxide concentrations. While many difficulties remain (such as the attribution of specific damages to a particular polluting activity or the differentiation between natural climate variability and anthropogenic climate change), a legal case applying existing legal principles becomes more likely as science advances.

At the same time, it should be noted that international enforcement mechanisms are still weak and developing and that it could be advisable to rely on multilateral negotiation rather than confrontational case-work to ensure compensation for climate damages is provided as well as capacity and finances for damage prevention (i.e., adaptation). A pre-Kyoto proposal by the Brazilian delegation has already put attribution of climate change to particular polluters/emitters on the agenda of the international climate regime (FCCC/AGBM/1997/MISC.1/Add.3; available at www.unfccc.int). Scientists and policy makers could intensify the discussion around this proposal to see whether it can form the basis for a practicable solution for residual climate damages in the future. Otherwise, a conciliation commission (Article 14 of the FCCC) might be able to find an equitable solution that cannot readily be agreed upon by Parties through negotiations.

5. Conclusions

Adaptation to current and future climate change is a cross-cutting issue with environmental and human dimensions. International, regional and domestic legal systems influence the adaptive capacity and adaptation choices of developing countries. Further work in natural sciences can help use the leverage of some legal provisions to ensure that timely and effective adaptation policies are implemented in developing countries. However, there are also potential conflicts that must be taken into account, and—in the best case—prevented. Thus, there is still abundant scope for lawyers and social scientists to develop a methodological structure in which adaptation to climate change can become an integrated feature of sustainable development. Also, the international climate regime—the pertinent forum for decision-making—has yet to develop coherent institutional structures and detailed rules to ensure the effective implementation of obligations concerned with adaptation. Against the background of increasingly disastrous impacts of climate change on economies and people in developing countries, this task must be taken seriously while keeping in mind that mitigation is still the best remedy against the worst impacts of anthropogenic climate change.

Endnotes

---

[1] Regimes are social institutions that define practices, assign roles and guide the interaction of occupants of such roles within given issue areas (see Young, 1994,

p. 3). In international law, a regime connotes the entirety of rules and practices associated with the management and implementation of one or several interrelated, international treaties.

[2] These decisions concern, inter alia, the operation of the protocol's flexible mechanisms and the compliance system; see FCCC/CP/2001/13/Add.2 and FCCC/CP/2001/13/Add.3, available at www.unfccc.int.

[3] This is recognized by all COP decisions regarding adaptation, in particular by the three-stage approach established at COP1, see FCCC/CP/1995/7/Add.1, available at www.unfccc.int.

[4] New guidelines for Non-Annex 1 National Communications were agreed at COP8. The document number was not available at the time of writing, but will be available in FCCC/SBI/2002/6/.

[5] A new "Least Developed Country Fund" was established primarily to fund NAPAs. At the 8th Conference of the Parties it was decided that the NAPA guidelines do not need any revision at this stage (FCCC/SBI/2002/L.8, available at www.unfccc.int), however, annotated guidelines were produced by the Least Developed Country Group, see document FCCC/SBI/2002/INF.14 at that site. Also, see Chapter 9 of this volume.

[6] Activities in this sense include projects, integration into other activities, capacity building and policy reform.

[7] ILM = International Legal Materials, UNTS = United Nations Treaty Series; both are available in law libraries. For an overview of the international human rights instruments, see Brownlie (1998). On the Inter American System, see Harris and Livingstone (1998). For an analysis of the relationship between human rights and environmental degradation, see Boyle and Anderson (1996).

[8] Article 4.1 (f) reads: "Parties shall . . . take climate change considerations into account, to the extent feasible, in their relevant social, economic and environmental policies and actions, and employ appropriate methods, for example impact assessments, formulated and determined nationally, with a view to minimizing adverse effects on the economy, on public health and on the quality of the environment, of projects or measures undertaken by them to mitigate or adapt to climate change".

[9] Residual damages occur when adaptation measures are not possible or are not carried out because of economic or technical constraints. It should be noted that at least under international law, states are not obliged to undertake all adaptation measures imaginable, but can themselves determine what is "adequate". Residual damages could thus occur to a great extent in one country but not in another.

[10] Mitigation (as opposed to adaptation) is the prevention of anthropogenic climate change at the source by either reducing greenhouse gas emissions or enhancing sink capacities (terrestrial or other). Mitigation is the focus of the UN FCCC and the Kyoto Protocol. In the framework of a legal analysis of responsibility for climate change damages, mitigation can be viewed as "indirect" damage prevention, concerned with actually preventing a risk of damages from anthropogenic climate change.

[11] In national legal systems, damage prevention is a duty upon the person responsible for future damage, be it either as an active obligation or only in the form of a duty to bear the costs for damage prevention incurred by the victim or third persons.

[12] International legal obligations can be found in either treaty law or customary international law. Customary International Law is developed by state practice and *opinio juris*, i.e., the perception of states that a certain behavior actually reflects a rule of international law.

Acknowledgments

The author wishes to thank Richard Klein (PIK) and Richard Tol (Professor for Global Change and Sustainability, University of Hamburg) for advice and support and for helping me understand the science of adaptation to climate change, Matthias Buck (Research Unit Environmental Law, University of Hamburg) for inspiring comments, and the German Scholarship Foundation (Studienstiftung des deutschen Volkes) for the financial support provided.

References

Bergkamp, N. and Orlando, B. (1999), Wetlands and Climate Change. Exploring Collaboration between the Convention on Wetlands and the UN Framework Convention on Climate Change. International Union for Conservation of Nature and Natural Resources, Geneva.

Bodansky, D. (1993), The UN Framework Convention on Climate Change: a commentary. *Yale Journal International Law*, **18**, 451.

Boyle, A. and Anderson, M. (1996), *Human Rights Approaches to Environmental Protection*. Clarendon Press, Oxford.

Brownlie, I. (1998), *Principles of Public International Law*. Clarendon Press, Oxford.

Callaway, J. et al. (1998), Adaptation costs: a framework and methods. In *Mitigation and Adaptation Costs Assessment, Concepts, Methods and Appropriate Use*, Francis, D., Brooke, C. and Halsnæs, K., eds. UN Environment Programme, New York.

Crawford, J., Peel, J. and Olleson, S. (2001), The ILC Articles on State Responsibility for Internationally Wrongful Act: Completion of the Second Reading. 13 *European Journal of International Law* (2001) 963.

Forest, C. E., Stone, P. H., Sokolov, A. P., Allen, M. R. and Webster, M. D. (2002), Quantifying uncertainties in climate system properties with the use of recent climate observations. *Science* **295**, 113-117.

Gibney, M., Tomasevski, K. and Vedsted-Hansen, J. (1999), Transnational state responsibility for violations of human rights. *Harvard Human Rights Journal* **12**, 267.

Grubb, M., Vrolijk, C. and Brack, D. (1999), *The Kyoto Protocol: A Guide and Assessment*. Royal Institute of International Affairs, London.

Harris, D. and Livingstone, S. (1998), *The Inter-American System on Human Rights*. Clarendon Press, Oxford, UK

IPCC (1994), *IPCC Technical Guidelines for Assessing Climate Change Impact and Adaptations. Part of the IPCC Special Report to the First Session of the Conference of Parties to the UN Framework Convention on Climate Change, Working Group II, Intergovernmental Panel on Climate Change*. University College London, UK and Center for Global Environmental Research, National Institute for Environmental Studies, Tsukuba, Japan.

Klein, R. (2001), Adaptation to Climate Change in German Official Development Assistance. Deutsche Gesellschaft für Technische Zusammenarbeit (GTZ). Available at www.gtz.de.

Klein, R. (In press), Adaptation to climate variability and change: what is optimal and appropriate? In *Climate Change and the Mediterranean: Socio-Economics of Impacts, Vulnerability and Adaptation*, Giupponi, C. and Schechter, M., eds. Edward Elgar, Cheltenham, UK.

Knutti, R., Stocker, T. F., Joos, F. and Plattner, G.-K. (2002), Constraints on radiative forcing and future climate change from observations and climate model ensembles. *Nature* **416**, 719-723.

McCarthy, J. J., Canziani, O. F., Leary, N. A., Dokken, D. J. and White, K. S., eds. (2001), *Climate Change 2001: Impacts, Adaptation, and Vulnerability. Contribution of Working Group II to the Third Assessment Report of the Intergovernmental Panel on Climate Change*. Cambridge University Press, Cambridge, UK.

Miles, E. L., Underdal, A., Andresen, S., Wettestad, J. Skodvin, T. and Carlin, E. M. (2001), *Environmental Regime Effectiveness: Confronting Theory with Evidence*. MIT Press, Cambridge, MA, USA.

Oberthür, S. and Ott, H. (1999), *The Kyoto Protocol*. Springer, Berlin.

Öko-Institut (2001), Requirements of Climate Protection with Regard to the Quality of Ecosystems: Use of Synergies between the Framework Convention of

Climate Change and the Convention on Biological Diversity. Report prepared on behalf of the Federal Environmental Agency of Germany, Research Project 200 41 204. Available at http://www.umweltbundesamt.de.

Perrez, S. (1996), The relationship between "permanent sovereignty" and the obligation not to cause transboundary environmental damage. *Environmental Law* **26**, 1187.

Protection of the Atmosphere: Statement of the Meeting of Legal and Policy Experts, Ottawa, 22 February 1989. Reprinted in *American University Journal for International Law and Policy* **5/2** (1990), 529-542.

Sands, P. (1992), The United Nations Framework Convention on Climate Change. *RECIEL* **1**, 270.

Sands, P., ed. (1995), *Principles of International Environmental Law, Volume I: Frameworks, Standards and Implementation; Volume II: Documents in International Law.* Manchester University Press, New York.

Smit, B., Burton, I., Klein R. J. T. and Wandel, J. (2000), An anatomy of adaptation to climate change and variability. *Climate Change* **45**, 223-251.

Sprinz, D. and Luterbacher, E., eds. (1997), International Relations and Global Climate Change. Available at www.pik-potsdam.de/dept/soc/e/reports/pr211.htm.

Stokke, F. (2001), The Interplay of International Regimes: Putting Effectiveness Theory to Work. FNI Report 14/2001. Available at www.fni.no.

Tol R. and Verheyen, R. (2001), Liability and Compensation for Climate Change Damages—A legal and economic assessment, working paper. Submitted to Energy Policy, currently being revised. Available at www.uni-hamburg.de/Wiss/FB/15/Sustainability/liability.pdf

Tomuschat, C. (1999), International law: ensuring the survival of mankind on the eve of a new century. *Recueil des Cours* **281**, 1.

UNEP/WMO (1992), *United Nations Framework Convention on Climate Change.* UN Environment Programme and World Meteorological Organization, Information Unit on Climate Change, Geneva. Available at www.unfccc.int.

UNFCCC (1997), *Kyoto Protocol to the UN Framework Convention on Climate Change*. Available at www.unfccc.int.

Verheyen, R. (1997), Der Beitrag des Völkerrechts zum Klimaschutz. In *Klimaschutz im Recht,* Koch, H.-J. and Caspar, J., eds. Nomos, Baden-Baden, Germany, 29.

Watson, R. T. and multiple authors (2001), Summary for policymakers. In *Climate Change 2001: Synthesis Report—Contribution of Working Groups I, II, and III to the Third Assessment Report of the Intergovernmental Panel on Climate Change.* Cambridge University Press, Cambridge, UK, 1-34.

Young, O. R. (1994), *International Governance: Protecting the Environment in a Stateless Society.* Cornell University Press, Ithaca, NY, USA.

9. Adaptation Funding—Legal and Institutional Issues

Roda Verheyen
Research Unit Environmental Law
University of Hamburg, Germany

The paper analyzes the legal framework for adaptation funding, focusing on the obligations contained in the UN Framework Convention on Climate Change and the Kyoto Protocol. It compares the results with the funding practice of the Global Environment Facility and identifies remaining problems and areas for future research.

1. Introduction

While it is still unclear what exactly the adaptation needs of developing countries are and what measures to increase adaptive capacity will cost, they represent an additional stress on the economies of developing countries. This paper discusses legal and practical issues relating to adaptation funding, focusing on financial flows from industrialized countries to developing countries within the framework of the international climate regime represented by the UN Framework Convention on Climate Change (FCCC) of 1992 (UNEP/WMO, 1992) and the Kyoto Protocol of 1997.[1] In this context, I draw on some definitions and assumptions made in Chapter 8 of this volume.

It should be borne in mind, however, that implementing and financing adaptation measures or measures to enhance the adaptive capacity of a given country or region would generally be the responsibility of the respective country with regard to both natural climate variability and anthropogenic climate change. In fact, many domestic legal systems provide a legal basis for arguing that either the state or other bodies of governance must bear the costs for adaptation measures necessary for the survival of communities and individuals or the protection of ecosystems. Those would correspond to rights of citizens or their natural resources to be protected from hazards by the state (see Chapter 8 of this volume). At the same time, private individuals or communities might, under national law, be responsible for preventing damage to the interests of others. Thus, any financial obligations of states toward other states must be seen in conjunction with the role of private entities within the recipient state in undertaking and financing adaptation measures.

In practice, funds are lacking in many developing countries even for the management of current climate variability, and in particular weather extremes. And also with respect to outside support, experience with droughts and floods shows that adaptation to current climate variability is not financed in a coherent manner, despite the effort of the United Nations through the International Strategy for Disaster Reduction[2] and the engagement of many development agencies. In many instances, efforts are still focused on disaster relief rather than preparedness or prevention (Chapter 4 of this volume; International Federation of Red Cross and Red Crescent, 2001). Moreover, there is no obligation on donors to finance specific measures with a view to adapting to climate change or, in fact, to provide any development assistance at all.

The "Right to Development" is still being discussed by developing nations as basis for mandatory development aid (see UN General Assembly Resolution A/RES/41/128, 4 December 1986),[3] and many international declarations and treaties emphasize the need for financial cooperation to achieve sustainable development (see the Monterrey Consensus, adopted at the International Conference on Financing for Development, 22 March 2002, available at www.un.org/esa/ffd as well as the World Summit for Sustainable Development—Plan of Implementation, available at www.un.org). One of the current development goals of the United Nations is the promotion of "environmentally sustainable development" (United Nations, World Bank, IMF, OECD; A better world for all—Towards the development goals, Geneva 2000).[4] But no binding international law rule exists on this issue, and thus all funds provided by developed nations for activities that could qualify as supporting adaptation to climate change are voluntary. In this respect, the climate regime represents a significant change both in legal and political terms.

It is important to note in this context that the general trend in development cooperation in the past decades has been to move away from official development aid (ODA) toward greater emphasis on the role of trade, private investment and the mobilization of domestic funds. This also explains why the official target of 0.7 percent of GDP annually for ODA (see, e.g., UN Millennium Declaration, UN General Assembly Resolution 55/2) has been met by a only small minority of industrialized countries—in 2000, for example, it was met by the Scandinavian countries, the Netherlands and Luxembourg, while the United States provided only 0.10 percent and Germany 0.27 percent of their respective GDP for development aid (OECD Doc. PAC/COM/NEWS(2002) 34). The results of the UN Conference on "Financing for Development" in March 2002 generally support this trend, even if a certain change might also be discernible—while stressing the importance of trade and private investment, the United States committed to increasing ODA levels by US$5 billion by 2005, and the European Union officially set an internal ODA target of 0.39 percent of GDP by 2006, which might mean an additional US$20 billion by 2006 (statements made by the United States, 19 March 2002, and the EU, 20 March 2002, at Monterrey, available at www.un.org/esa/ffd). These targets were confirmed at the so called Rio + 10 summit in Johannesburg in August 2002 (WSSD, see above). Yet, the 0.7 percent target has not been placed into a firm program with timetables, and neither have industrialized countries committed to an overall funding figure.

In the framework of environmental treaties however, providing additional financial resources to developing countries for implementation (and indeed, participation) purposes has become a general practice since the late 1980s. Accordingly, during the negotiations of the FCCC, developing countries emphasized that the implementation of any obligations by them, including the duty to implement adaptation measures, would depend on new financial commitments by the industrialized countries (see Article 4.7 of the FCCC). Today, it seems clear that such new resources must be coordinated with existing development and disaster relief funding to ensure efficient responses to both natural and anthropogenic climate change. The following section sketches the funding obligations, legal problems and uncertainties as well as the current funding practice for adaptation activities, focusing on the international climate regime.

2. Convention Funding

The Convention contains two categories of financial obligations, each of which apply only to Parties listed in Annex II to the FCCC: those financial obligations generally aimed at helping developing countries implement their duties under the Convention (Article 4.3), and those obligations targeting adaptation specifically (Article 4.4). While the Convention does not require a requisite level of financial resources, the financial obligations in Articles 4.3 and 4.4 are mandatory.

Recipients of FCCC-related funding are developing states not included in Annex I to the Convention (which lists developed state Parties).[5] These non-Annex I countries comprise a group of states at very different stages of economic development. This fact has already caused problems. Because of their better starting conditions to apply for funding, larger developing countries, such as China and India, have been able to secure a majority of funding provided under the Convention, leaving especially least developed countries (LDCs) angered by the lack of implementation of financial obligations designed for their benefits. At the same time, the per capita contributions to these large countries are rather small, compared to support provided to less populated least developed countries.

Pursuant to Article 21.3 FCCC and several decisions of the Conference of the Parties (COP),[6] Convention funding is provided via the Convention's financial mechanism, the Global Environment Facility (GEF), and industrialized states are encouraged to provide bilateral or other types of funding. The GEF was established as a joint project of UNEP, UNDP and the World Bank in 1991 to address the financial needs to fight biodiversity loss, climate change, degradation of international waters and ozone depletion—the four focal areas.[7] Industrialized countries pledge contributions to the GEF in regular intervals, thereby "replenishing" the GEF trust fund. The third replenishment of the GEF was, after much delay, successfully concluded in August 2002.[8] In addition, at the Johannesburg summit some EU countries pledged to add further $80 Mio, thus reaching the initial goal of $3 billion. The GEF has an operational strategy, upon which all its decisions are based (Global Environment Facility, Operational Strategy, February 1996, http://gefweb.org). It stipulates, *inter alia*, that the projects financed or cofinanced by the GEF shall result in "global benefits". In the view of the GEF, global benefits are obtained whenever a global environmental objective is met. In the context of adaptation funding, this basic principle should be revised and removed from the operational strategy because adaptation often benefits only the region or country in which measures are undertaken.

At the time of the establishment of the GEF, its climate change projects focused primarily on mitigation, organized into four areas (removing barriers to energy efficiency and energy conservation; promoting the adoption of renewable energy by removing barriers and reducing implementation costs; reducing the long-term costs of low greenhouse gas emitting energy technologies; and supporting the development of sustainable transport); however, the GEF is now also entrusted with the administration of the new Kyoto Protocol Adaptation Fund (Adaptation Fund), the Special Climate Change Fund (SCC Fund), and the Least Developed Country Fund (LDC Fund), which were each established by the Parties at the second session of COP6 (COP6*bis*) held in Bonn in July 2001 (Decision 5/CP.6, FCCC/CP/2001/5, this was confirmed at COP7, see Decision 7/CP.7 and 10/CP.7 in FCCC/CP/2001/13/Add.1, available at www.unfccc.int). These new funds are explicitly foreseen by the COP to provide additional funding to meet the commitments under (inter alia) Articles 4.3 and 4.4 of the FCCC. It is thus clear that the GEF climate funding practice will have to change substantially.

Articles 4.3 and 4.4. of the FCCC

Under Article 4.3, industrialized countries (Annex II) are obliged to "provide new and additional financial resources to meet the agreed full costs" of developing countries' reporting obligations (under Article 12.1 FCCC). Second, they "shall also provide such financial resources . . . needed by the developing country Parties to meet the agreed full incremental costs of implementing measures that are covered by paragraph 1". This includes, inter alia, the program and adaptation measures in Article 4.1 (b). Even a decade after the adoption of the Convention, the interpretation and application of terms such as "incremental costs" and "resources . . . needed" are ambiguous.

Incremental Costs

The incremental cost concept (which, incidentally, is not specified in the climate treaties) is linked to the fact that the FCCC applies only to anthropogenic climate change and, thus, Convention funding is not meant to finance all sorts of projects in developing countries. Moreover, the GEF is committed to fund only "agreed incremental costs of measures to achieve global environmental benefits" (Instrument for the Establishment of the Restructured GEF, 1994, available at gefweb.org; see Werksman, 1993, and Bodansky, 1993). Therefore, only measures to adapt to climate change caused by human activities and measures that would not be undertaken otherwise should be financed by the GEF from Convention funds. In other words, the GEF will dedicate its funds "to achieving global environmental benefits rather than to achieving development and local environmental benefits". Thus, if a measure supports other aims or would be implemented anyway with funding from other sources, the costs are—in the strict sense—not incremental (see GEF/C.7/Inf.5, Incremental Costs, February 29, 1996, available at www.gefweb.org).

The guidelines provided by the GEF for the assessment of incremental costs have been recognized by some to be inadequate for adaptation purposes, yet the GEF Council decision of 2002 on the new funds reiterates that "resources from the new funds shall be used to meet the agreed full incremental costs of implementing measures" (GEF/C.19/6, see also the reiteration of the concept in the Beijing Declaration, above). The yardstick for adaptation can logically only be "costs induced by climate change" (or the like). As will be outlined below, the COP decisions on funding for adaptation measures does not refer to the concept explicitly. However, the donor community has already highlighted that they wish to keep the concept of incremental funding for the GEF trust fund, which means that the trust fund (as opposed to the new funds) cannot and will not fully fund adaptation measures.

At the same time, the text of Article 4.3 implies that financial entitlement for adaptation purposes is to be determined by real needs in developing countries ("resources needed"). Developing countries would thus be well advised to develop assessments of their needs and concrete programs to meet those, which could for the basis of funding claims. The right to claim adaptation expenses is, however, qualified by the requirement to *agree* on costs with the GEF (Article 4.3, first and second sentences).

Particularly Vulnerable Parties

Article 4.4 of the FCCC stipulates that Annex II Parties "*shall* also *assist* Developing Country Parties that are *particularly vulnerable* to the adverse effects of climate change in meeting *costs of adaptation* to those adverse effects" (emphasis added). This provision was introduced by the Alliance of Small Island States (AOSIS) in the last days of the negotiations and is a result of political treaty language. The provision's mandatory character was opposed by the United States and others because it was understood to provide an unlimited basis for funding (Report of the work of the INC, Geneva, 20 December 1991, A/AC.237/15, at 41, section IV, para. 2.2(a).) Therefore, some have interpreted Article 4.4 as the most costly provision of the FCCC and that Article 4.4 "amounts to an implicit acceptance of responsibility for causing climate change" (Sands, 1992). However, the precise legal content of Article 4.4 is difficult to ascertain.

The Convention does not define the term "particularly vulnerable", which is set out in Article 4.4. The preamble of the FCCC states that "low lying or small island countries, countries with low-lying coastal, arid and semi-arid areas or areas liable to floods drought and desertification, and developing countries with fragile mountainous ecosystems are particularly vulnerable to the adverse effects of climate change". However, the preamble does not refer explicitly to developing countries and the text of the preamble can be used only as an interpretative aid and not as an authoritative rule (Vienna Convention on the Law of Treaties, 8 I.L.M, 1969, 679, Article 31; see also Yamin, 1998).

In this context, the list of countries with "specific needs and concerns" contained in Article 4.8 of the Convention could be helpful. This list includes several categories of countries that are not incorporated in the preamble text, such as countries with forested areas and areas liable to forest decay; countries prone to natural disasters; countries with areas of high urban atmospheric pollution; countries with fragile ecosystems; countries whose economies are highly dependent on income generated from the production, processing and export and/or consumption of fossil fuels; and landlocked and transit countries.

However, it does not solve the problem of interpreting the phrase "particularly vulnerable", since it does not refer to the term. Moreover, applying the list to Article 4.4 would lead to virtually all developing countries being able to claim adaptation costs via Article 4.4. This result would blur the difference between Articles 4.3 and 4.4 of the FCCC.

Yamin (1998) suggests that a list of vulnerable countries should be negotiated. Citing various international declarations, she argues that the international community has accepted that small island developing states are particularly vulnerable. This could be questioned, since the Convention itself emphasizes the needs of Africa as a region in Article 4.1(e). However, because Article 4.1(e) is covered by the funding provision in Article 4.3, the comparison does not lead to a better understanding of which countries are "particularly vulnerable" under the auspices of Article 4.4.

While the text is inconclusive, Parties have agreed that it is necessary to identify particularly vulnerable Parties in order to implement Article 4.4 (FCCC/CP/1995/7/Add.1, para. 1(d)). However, while the vulnerability of countries can be assessed scientifically, it is likely that most developing countries would qualify generally as "particularly vulnerable" because of their financial inabilities to protect themselves, bear the financial risk of catastrophic weather events and to mitigate the adverse effects of climate change. Against this background, it has been suggested that eligible countries could be ranked on the basis of vulnerability indices,[9] and that the IPCC should be entrusted with developing such an index, by building on its work on methods for vulnerability assessments. At the same time, such indices have been heavily criticized because the choice of criteria for prioritizing cannot be objective and thus, such indices are of very limited value for decision-making. It is thus questionable whether a just and negotiable prioritization can be (and should be?) achieved at all.[10]

"Assist in Meeting the Costs"

To add to this seemingly unsolvable problem, Article 4.4 remains ambiguous about the extent of funding to be provided. The use of the phrase "to assist . . . in meeting costs of adaptation" in Article 4.4 as opposed to the phrase in Article 4.3 that "the agreed . . . costs" must be covered implies that not all costs of adaptation would need to be borne by Annex II Parties. Unfortunately, Article 4.4 has not been discussed extensively in the climate change negotiations, and thus, no clear views have been offered on this issue by the Parties. Importantly, however, and in contrast to Article 4.3, it does not refer explicitly to the concept of incrementality.

The starting point for any interpretation of Article 4.4 should be the adaptation needs of countries rather than the willingness of Annex II countries to pay for adaptation measures. This principle flows both from Article 4.1(b) (see Chapter 8 of this volume) and from the open formulation used in Article 4.4. If Annex II Parties must "assist" in meeting the costs of adaptation, clearly, the extent of assistance needed should be measured against the real needs of developing countries. Moreover, Parties are not only to provide the additional costs needed to implement a project/measure (incremental costs). Therefore, scientific methods for assessing these needs play a major role in the implementation of both Article 4.3 (second sentence) and Article 4.4. Such an assessment process was started at COP 1 in 1995 with its first decision on guidance for the financial mechanism, which provides that the GEF should—until further notice—finance vulnerability assessments to determine adaptation needs. This is discussed below in connection with adaptation funding in practice. Such an assessment process is also at the heart of the National Adaptation Programmes of Action (NAPAs) (see Chapter 8 of this volume).

3. Kyoto Protocol Funding

Funding for adaptation under the Kyoto Protocol is linked to the operation of the Protocol's Clean Development Mechanism (CDM, Article 12 of the Kyoto Protocol). The CDM is a project-based mechanism, designed to enable climate-friendly investments in developing countries. To meet their Kyoto reduction or stabilization targets, public and private entities in developed countries listed in Annex I of the Convention can undertake CDM projects in developing countries that will result in emission reductions. Emission reductions must be certified by an independent administrative body (designated operational entity) before credit for those reductions is given. After emissions reductions have been certified, certified emission reductions (CERs) credits will be issued and transferred to the investor or investor country. CERs can be used by Annex I Parties toward meeting their Kyoto emissions reduction targets; for the CDM and its history, see Werksman (1998) and Oberthür and Ott (1999).

Article 12.8 of the Kyoto Protocol provides that "a share of proceeds" from CDM project activities shall be used to "assist developing country Parties that are particularly vulnerable to the adverse effects of climate change to meet the costs of adaptation", as well as to meet the administrative costs of the CDM. At COP6bis and COP7, Parties agreed that the share of proceeds shall be "2% of the certified emissions reductions issued", and that CDM project activities in least developed country Parties shall be exempt from this obligation (Bonn Agreement, Para VI(1)(10); FCCC/CP/2001/CRP.11; and Marrakech Accords, Decision 17/CP.7, FCCC/CP/2001/13/Add.4, para. 15(a), available at www.unfccc.int). CDM project activities in LDCs are exempt from this obligation. These monies will be transferred to the newly established Kyoto Protocol, through the GEF, which will administer the fund. Only Parties to the Kyoto Protocol are eligible to receiving finance through the fund (Decision 10/CP.7, FCCC/CP/2001/13/Add.1, available at www.unfccc.int).

The term "particularly vulnerable" is not defined in the context of the Kyoto Protocol or in any of the related COP decisions. Thus, just as in the case of Article 4.4 of the Convention, the GEF does not have any clear guidance on which countries are eligible recipients for funding. The GEF must prioritize the needs of LDCs and, in particular, the needs of Small Island Developing States in its granting of funding; however, this guidance is not an authoritative interpretation of the term "particularly vulnerable". Moreover, it is uncertain to what extent CDM projects will be undertaken at all. The value of CDM credits is not predetermined, but will depend on market supply and demand, thus the level of funding that this fund will be able to provide remains to be seen. It is unclear whether the Adaptation Fund will receive CERs as such and be free to sell them whenever funding is being sought or whether it will receive the monetary value of the CER after they are issued. However, in contrast to Articles 4.3 and 4.4 of the Convention, Article 12.8 of the Protocol ensures revenues for adaptation purposes as soon as the CDM starts operating (from 2000 onward). Article 12.10 of the Kyoto Protocol allows for the "prompt start" of the CDM. Projects can be undertaken before the Protocol enters into force and any generated CDM credits can be used for complying with the Kyoto targets. However, because it is currently uncertain when the Protocol will enter into force, it is unlikely that any distribution of funds from this source will happen before 2003 or 2004.

4. Some Conclusions

On the basis of this analysis of the legal situation with regard to adaptation funding, it is currently impossible to determine what exactly has to be financed by Annex II countries under the provisions of the Convention and the Kyoto Protocol.

First, following a literal interpretation of Articles 4.3 and 4.4 of the FCCC as well as Article 12.8 of the Kyoto Protocol, only adaptation to anthropogenic climate change must be financed, not adaptation to climate variability. As of yet, there is no method to separate the costs from these two sources of damage. However, recent COP decisions seem to abandon this principle. For instance, the guidelines for National Adaptation Programs of

Action (NAPA) adopted at COP7 (see Chapter 8) constitutes a clear deviation from the provisions of the FCCC in this respect as it permits funding for adaptation to both natural and anthropogenic climate change. However, this design decision taken in the framework of NAPAs will not solve future conflicts when funding for activities other than assessments and strategic planning is at stake.

Second, the "incremental costs" concept must be reevaluated. Many measures aimed at enhancing the adaptive capacity of developing countries would also qualify as general development measures and thus fall within the scope of the Convention only inasmuch as they are aimed specifically at adaptation, following the incremental cost yardstick in Article 4.3 of the FCCC. Article 4.4 does not explicitly refer to incremental costs, but since the GEF is committed to this concept as a whole, the concept seems to apply to adaptation funding as a whole. But how would one judge if a measure were fully aimed at adaptation or also at some other societal goals? And if only part of a measure would fall within the scope of the climate regime funding, how could the remainder be financed? The COP decisions already recognize the fact that adaptation funding will not achieve global environmental benefits (see further below); thus, the methodology for the incremental costs concept applied by the GEF, which is based on the "global environmental benefit" paradigm, cannot be used.

Third, "adequate adaptation" is an open term whose interpretation depends on the particular country and situation. Therefore, there is not yet any indication of what will be needed in any particular country and what is required overall. Thus, while in theory the provisions of the Convention are need-driven, there have been few attempts to actually define the needs that have to be met by Annex II Parties. This practice also contradicts the clear mandate in Article 11.3(d) of the FCCC on the financial mechanism, which calls on Parties to determine "in a predictable and identifiable manner the amount of funding necessary and available for the implementation of this convention and the conditions under which that amount shall be periodically reviewed." Interestingly, at COP8 in November 2002, developing country Parties attempted to establish a process for the joint determination of "aggregate funding requirements" by the COP and the GEF Council (see Draft decision "Financial Mechanism—Review of the Financial Mechanism", SBI Agenda item 5c), 28 October 2002, available from author), but failed because donor countries were taken by surprise by this move. But the decision that was adopted at COP8 at least requests the FCCC secretariat to prepare a report on the "determination of funding necessary and available for the implementation of the Convention" (Draft decision -/CP.8, Review of the financial mechanism, available at www.unfccc.int), i.e., on ideas about the operationalization of Article 11.3(d) of the FCCC.

Fourth, there have not been any credible attempt to prioritize target countries. While vulnerability assessments will identify vulnerable countries generally, the use of vulnerability indices is debatable and, thus, it is uncertain whether the implementation of Article 4.4 of the FCCC and the distribution of funds generated through Article 12.8 of the Kyoto Protocol (share of proceeds) can be supported by any prioritization at all. At the same time, the GEF's current first come, first serve practice (see below) must be replaced by a rational system to ensure the equitable and effective distribution of funds and to ensure that funding practices abide by the rules set out in the climate regime.

All of these issues need further exploration, both from adaptation experts and political scientists to produce results that can operate within the climate regime.

5. The GEF Funding Practice

These legal difficulties and uncertainties are currently managed through the funding practice.

Adaptation (and other) funding is provided to developing countries by the GEF, serving as the financial mechanism under the Convention (Article 11 of the FCCC and Decision 3/CP.4, FCCC/CP/1998/16/Add.1, at 8), and to a lesser extent to countries with economics in transition. Just as the other convention bodies that have entrusted the GEF with the administration of their respective financial obligations (such as the Convention on Biological Diversity and the UN Convention to Combat Desertification), the Parties to the FCCC provide guidance to the GEF in the form of COP decisions. Together with the general operational policy of the GEF, these form the basis for financing adaptation activities.

COP Guidance to the GEF

The basic rules on adaptation funding for the GEF are contained in the various decisions, above all the COP1 decision on the three-stage process for adaptation measures and planning (FCCC/CP/1995/7/Add.1)[11]:

- Stage I: Planning, which includes studies of possible impacts of climate change, to identify particularly vulnerable countries or regions and policy options for adaptation and appropriate capacity-building.
- Stage II: Measures, including further capacity-building, which may be taken to prepare for adaptation, as envisaged by Article 4.1(e).
- Stage III: Measures to facilitate adequate adaptation, including insurance, and other adaptation measures as envisaged by Article 4.1(b) and 4.4.

Stages II and III were envisaged for the long term for the particularly vulnerable countries or regions identified in Stage I.

Since COP decisions complement the provisions of the Convention, this three-stage approach constitutes a *de facto* limitation of activities to be funded in accordance with Articles 4.3 and 4.4 of the Convention in the short and medium term. Funding for the implementation of Articles 4.1(b) and 4.4 is not authorized until Stage III is reached. To fund Stages II and III, the COP is given a mandate to "decide on channels" of funding, indicating that they are unlikely to be financed via the general GEF trust fund. The GEF trust fund receives the contributions from developed states and finances projects and activities in all GEF focal areas, not only for climate change. Rather, the establishment of the new funds at COP6*bis* shows that many Parties wish to separate adaptation funding from the general climate core area. Through the various COP6*bis* decisions on funding issues, Annex II Parties also maintained that Stage III funding would not be provided in the short term at all.

The "enabling activities" under Stage I, are closely connected with reporting requirements under the Convention. Pursuant to the first sentence in Article 4.3 of the FCCC, Annex II countries must meet the full costs for national communication of Non-Annex I countries. The development of these fit the criteria of Stage I activities. Thus, in fact, Stage I activities meet only the obligation of the first sentence and do not implement the second sentence or Article 4.4. It is important to note that, because the GEF had to bear the full costs of all such activities, there has been no need for prioritization of requests.

The GEF has funded numerous enabling activities, such as vulnerability assessments in agriculture, forestry, coastal zones, water resources, health and natural ecosystems as components of national communications (GEF, 2000; GEF, 2002). Since Stage I activities are envisaged to identify "particularly vulnerable countries", and help to define this term in both

Article 4.4 of the FCCC and 12.8 of the Kyoto Protocol, they are a first step toward the implementation of those obligations. These assessments are, however, still not concluded in all developing countries. The recent GEF (2000) review notes the need to further prioritize work on vulnerability assessment and recognizes the lack of detailed information concerning the impact of climate change on water resources, food security and other sectors in developing countries. Stage I activities are also meant to identify "adequate" adaptation options, which could be essential to define the term "adequate" in Article 4.1(b) of the FCCC (see Chapter 8).

The restriction of funding to Stage I activities, possibly not covered by the provisions of the Convention, was criticized by many developing countries, especially small island developing states. Whether decisions of the COP can override, amend or supplement the treaty text is a contentious issue of public international law and cannot be explored here. To address these concerns, Parties at COP4 decided that the GEF should fund "adaptation response measures (Stage II activities) in particularly vulnerable countries and regions identified in Stage I activities (Decision 2/CP.4, FCCC/CP/1998/16/Add.1, para. 1(a), http://www.unfccc.int). Despite the COP4 decision and the identification of vulnerable countries and regions in Stage I projects, no additional funds were allocated to the GEF to fund Stage II activities. This was partially due to the fact that Stage II activities were not specified beyond the COP1 decision and it was thus difficult to identify eligible projects (Klein, 2001). Still, the GEF has since provided funding to several Stage II projects from the GEF's general trust fund. These have been provided on a first-come-first-serve basis, which, due to the limited number of applications, did not create difficulties.[12] Also, as the activities were qualified as enabling activities in the context of the development of National Communications, they are funded on the basis of the agreed full-cost principle.

After much discussion, particularly about adaptation funding, additional guidance was provided to the GEF at COP7 (Decision 6/CP.7, FCCC/CP/2001/13/Add.1, Additional guidance to an operating entity of the financial mechanism). The new guidance gives the following responsibilities to the GEF:

- The GEF is to support Stage II adaptation projects in regions identified as particularly vulnerable in Stage I assessments.
- The GEF should establish adaptation pilot or demonstration projects in regions that have undergone Stage I and II measures to show how

the assessments and plans can be translated into concrete projects that will provide real benefits and may be integrated into national policy and sustainable development planning.

- The GEF is to provide funding for capacity building in the areas of disaster preparedness and contingency planning for natural disasters such as droughts and floods in areas prone to extreme weather events.
- The GEF should set up or strengthen existing early warning systems for extreme weather events.

Within this framework, the GEF must prioritize the needs of LDCs and, in particular, the needs of small island developing states. However, as stated above, this emphasis cannot be seen as an authoritative interpretation of the term "particularly vulnerable" in Article 4.4 of the FCCC or relating to the Adaptation Fund (Article 12.8 of the Kyoto Protocol).

Also, the guidance does not enable funding for Stage III adaptation measures, which would include, *inter alia*, insurance measures and measures to implement Articles 4.1(b) and 4.4. This means that Stage III funding will not be made available. It will, as mentioned above, force the GEF to revise its operational strategy because of the new language, which requires projects to result in "real benefits" rather than in "global benefits". This problem is closely related to the definition or application of the incremental costs concept. However, even if the concept is not applied to adaptation measures, because the GEF operational strategy constitutes the overall framework of GEF's work, in my view, the strategy still needs to be clarified. The guidance also acknowledges the linkages between current weather extremes and future adaptation needs, since it allows countries to apply for funding for disaster preparedness regardless of whether such disasters are caused by anthropogenic climate change.

Another COP6bis decision (implementation of Articles 4.8 and 4.9 of the FCCC) complements the new guidance and specifies activities that should be supported by the GEF (Decision 5/CP.7, FCCC/CP/2001/13/Add.1). This list does not introduce new areas, but further stresses the need for vulnerability assessments and disaster preparedness. The list includes strengthening or establishing observation and monitoring networks (sea level rise, etc.) as well as research programs on climate variability and climate change; supporting enabling activities for vulnerability and adaptation assessment; promoting the transfer of adaptation technologies; technical training for climate change impact and vulnerability and adaptation assessments; supporting capacity building for preventive measures, planning and preparedness of disasters. It also emphasizes the need for transfer of adaptation technology, to implement the technology transfer obligation in Article 4.5 FCCC.

To enable the quick allocation of funds for NAPAs, the Parties at COP7 issued initial guidance to the GEF relating to the LDC Fund (Decision 27/CP.7, FCCC/CP/2001/13/Add.4, http://www.unfccc.int). Consequently, the GEF will meet the full agreed costs of preparing NAPAs, ensure complementary funding between the LDC Fund and the other Kyoto funds (Adaptation Fund, SCC Fund and the regular GEF climate trust fund), ensure the separation of the LDC Fund from the other funds administered by the GEF, and encourage the use of national or regional experts in preparing NAPAs. New footnote: Parties at COP8 decided that the NAPA guidelines would not need revision (see Chapter 8) and called on the GEF to "to ensure the speedy release and disbursement of funds" from the LDC fund for their preparation, see Draft decision --/CP.8, Guidance to an entity entrusted with the operation of the financial mechanism of the Convention, for the Least Developed Countries Fund, http://www.unfccc.int)

The separate decision on the SCC Fund and the Adaptation Fund makes it clear that dividing lines between the funds are entirely blurred. Both funds are meant to finance (Decision 5/CP.7, FCCC/CP/2001/13/Add.1, available at www.unfccc.int):

- adaptation activities where sufficient information is available to warrant such activities, inter alia, in the areas of water resources management, land management, agriculture, health, infrastructure development, fragile ecosystems, and integrated coastal zone management
- improved monitoring of diseases and vectors affected by climate change, and related forecasting and early-warning systems, and in this context improving disease control and prevention
- capacity-building, including institutional capacity, for preventive measures, planning, preparedness and management of disasters relating to climate change, including contingency planning, in particular, for droughts and floods in areas prone to extreme weather events
- strengthening existing and, where needed, establishing national and regional centers and information networks for rapid response to extreme weather events, utilizing information technology as much as possible.

The Kyoto Adaptation Fund shall also finance "concrete adaptation projects and programmes in developing country Parties that are Parties to the Protocol", as well as activities listed in the decision about the implementation of Article 4.8 FCCC listed above.

Assessments and needs identification are at the core of these new guidance documents. Only the Adaptation Fund seems to have the mandate to fund actual adaptation projects and measures to enhance adaptive capacity of developing country Parties. To what extent the costs of such projects are to be borne by the fund remains unclear.

Parties at COP8 discussed at length how the three new funds could be made operational without creating overlaps and inefficiencies. Since the Adaptation Fund will start operating only once the Kyoto Protocol is in force, no detailed discussion took place about its guiding elements. On the SCCF, however, developing countries submitted a list of activities to be funded (both in the area of adaptation and on the issue of impacts of response measures, i.e., impacts of climate protection measures on developing country economies) on a full cost or incremental cost basis "as appropriate", a concept that was not acceptable to donor states (Draft decision on the SCCF, 28 October 2002, available from author). Finally, Parties agreed to initiate a process with a view to providing further guidance to the GEF, to adopt such guidance at COP 9 in 2003 (Draft decision --/CP.8, Initial guidance to an entity entrusted with the operation of the financial mechanism of the Convention, for the operation of the Special Climate Change Fund, available at www.unfccc.int).

Principles of GEF Adaptation Funding

Overall, the new regulations and guidelines put a new emphasis on adaptation within the climate regime. While the relationships between the listed measures and categories, target countries, real financial needs, and the institutional set-up remain confusing, there are clearly some principles that will guide adaptation activities and funding within the climate regime. These principles flow from both the convention text and the multitude of COP decisions:

- "Macro" measures to enhance adaptive capacity are eligible for funding, as well as "micro" adaptation projects.
- To qualify for funding under the FCCC, adaptation to anthropogenic climate change must not necessarily be separated from adaptation to natural climate variability.

- Past experiences from coping with, for example, extreme weather events must be taken into account to help to achieve the goals of the Convention.
- The institutional set-up for supporting decision-making and planning processes in developing countries must be improved to integrate adaptation goals into general policy planning. A first step is the newly established expert group on LDCs established by the COP in conjunction with the LDC Fund, comprising 12 experts on adaptation and development matters, which shall provide advice on the preparation and implementation for NAPAs, as well as on capacity building needs for LDCs in support of the preparation and implementation of NAPAs, while implementation responsibility lies with independent country experts (Decision 29/CP.7, FCCC/CP/2001/13/Add.4, see also the LDC Group's input on NAPAs, FCCC/SBI/2002/Inf.14, both available at www.unfcc.int).
- Enhancing adaptive capacity and implementing adaptation are inextricably linked with development. The donor community has indicated that it does not wish to fund "stand-alone adaptation projects". While donors should not maybe predetermine funding decisions, it is true that adaptation needs must be integrated with other policy and development planning processes. Some of the main tasks of the LDC expert group will be to coordinate NAPAs with other activities in LDCs relevant to adaptation to climate change as well as mainstreaming NAPAs into regular development planning. NAPAs are also meant to build on existing plans and programs (such as national action plans under the UN Convention to Combat Desertification, and national biodiversity strategies and action plans under the Convention on Biological Diversity). Parties should ensure that such work is not only restricted to LDCs but also undertaken for all developing country parties. Those could make use of the full funding provided by the GEF for national communication development, in which such integrative assessments could feature.
- Cost-effectiveness is only one criterion for decision-making in the face of multiple adaptation choices (see Chapter 8 of this volume).
- Adaptation choices must be country-driven; for instance, each country must decide what it perceives as adequate adaptation. An experience from the past 10 years of Convention history is that developing countries have often not been interested in developing vulnerability

assessments (as part of national communications or otherwise). There is a clear need to assess why this was the case and how this problem can be remedied in the future. Without ownership by a country, adaptation projects will not be planned, designed or undertaken at all.

6. Remaining Problems

While the increased focus on adaptation funding and the relatively detailed guidance texts have resulted in establishment of the above-mentioned principles for adaptation funding, many problems remain. These problems can be summarized into eight distinct issues.

Predictability of Funds
There is no predictability of funds for adaptation or enhancing adaptive capacity in developing countries, even after the impressive political commitments made in Bonn at COP6*bis*. The European Union, Canada, Iceland, Norway, New Zealand and Switzerland, for instance, pledged 450 million Euros per year by 2005 (to be reviewed in 2008) for developing countries at COP6*bis* (FCCC/CP/2001/5/Add.1, at 44, http://www.unfccc.int). This sum includes contributions to the GEF generally (including mitigation); bilateral and multilateral funding additional to current levels; funding for the SCC Fund, the Adaptation Fund, the LDC Fund; and funding derived from the share of proceeds of the CDM following entry into force of the Kyoto Protocol. It is unclear how these monies will be allocated, even after further discussion at COP8.

Additional Funding
While the COP7 decision on funding for developing countries stipulates that funding should be "additional", "predictable" and "adequate", no mechanisms have been put in place to ensure that such requirements are met. Current practice also runs short of the mandate in Article 11.3(d) of the FCCC to determine the amount of funding available in a "predictable and identifiable manner".

As mentioned above, a process was put in place at COP8 to work on this issue within the FCCC framework. Funding could be made more predictable with the adoption of a timetable, which could include minimum amounts by donor country and general areas of allocation (for instance, technology transfer, capacity building, adaptation) covering the subsequent three years and updated annually, so that planning and programs can be developed over longer time periods. Such programming and long-term financial commitments are particularly necessary to engage in measures aimed at enhancing adaptive capacity because the duration and costs of such activities might be much less predictable than micro projects. If solutions in this area are not swiftly negotiated, the conciliation mechanism of the Convention could be engaged by developing countries to ensure compliance with Articles 4.3 and 4.4 of the FCCC. Article 14.7 of the FCCC states that "additional Procedures relating to conciliation shall be adopted by the Conference of the Parties as soon as practicable, in an annex on conciliation". To date, this annex has not been adopted.

In addition, the OECD Development Assistance Committee (DAC) could become involved. It receives annual reports on ODA from all donors and therefore has an overview of country activities. Annex II Parties could instruct the DAC to prepare an "additionality report", to be combined with the annual funding reports that Annex II Parties are required to present to the COP for review.[13]

Burden Sharing

The Convention does not stipulate a formula for burden-sharing between Annex II countries. So far, countries contribute to the GEF trust fund at their own discretion, and there are no objective criteria for defining any particular donor's obligations.

This is an issue that must be addressed at some stage. Experiences from the Montreal Protocol and general UN burden-sharing practices can serve as guidelines for the COP to develop a viable method.[14] The principle of common but differentiated responsibility (Article 3.1 of the FCCC) could also guide the negotiations, maybe leading to a distribution of funding obligations on the basis of emission shares or per capita emission shares, combined with economic indicators.

The New Funds

It is still unclear how the new funds will function and how they will relate to the general GEF trust fund. The GEF Council decision on "Arrangements for the Establishment of the New Climate Change Funds" (GEF/C.19/6) stipulates that the GEF trust fund will be kept separate from the three new funds. However, the activities to be funded through the special climate change fund and the LDC Fund in particular overlap with what has been funded by the GEF trust fund (enabling activities). Only the LDC Fund has a relatively concrete objective (to fund the NAPAs), and for this fund, the GEF has (together with experts) already developed operational guidelines to offer support to interested LDCs quickly. For the other funds, the COP will develop new guidance. However, the GEF's intent to develop, following the new COP guidance, an assessment of financing needs for the "identified priorities" is promising.

New Approaches

The GEF funding practices in the climate arena are not easily transferable to activities aimed at enhancing adaptive capacity or adaptation projects.

Since 1991, approximately US$1.3 billion in grants has been made available from the GEF trust fund for climate change projects, with renewable energy and energy efficiency projects dominating the portfolio. Adaptation projects have been supported only through enabling activities as described above. The GEF's reports to the COP indicate that most of the funding in this area has been allocated to the Caribbean and Pacific Islands programs. The kinds of adaptation project proposals the GEF is likely to receive, especially under the Adaptation Fund, will differ greatly from these ongoing programs. Adaptation will force the GEF and the implementing agencies such as UNEP into integrative and innovative screening of projects and implementation of project cycles. Closer ties between the biodiversity focal area and the climate change focal area are already planned to ensure that adaptation activities are coordinated with the aims and objectives of the GEF for biodiversity conservation. This is a first step, which must be followed by methods that will allow the GEF and other donors to assess and approve adaptation projects effectively. Such methods also could then be used for bilateral official development assistance. The development of a GEF Adaptation Strategy also seems a step forward, as well as the inclusion of projects aimed at adaptation in other focal areas in the GEF's assessment of its activities for the COP (see GEF, 2002).

Also, the incremental cost issue (application to adaptation) must be addressed and solved.

Who Comes First?

The GEF's current first-come, first-serve practice—for the implementation of Articles 4.4 of the FCCC and 12.8 of the Kyoto Protocol—does not comply with the provisions of the FCCC and the Protocol ("particularly vulnerable") and will become inappropriate as requests for adaptation funding increase in size and number. The COP must provide the GEF with guidelines based on sound science to clarify how to prioritize funding requests. This may be achieved by institutionalizing such decisions within the climate regime by establishing (among other things) an expert group (similar to the LDC expert group) whose mandate is to provide further detailed and flexible case-to-case guidance to the GEF. On the other hand, many would argue that the GEF itself is the best institution for developing such advice and that further institutions lead only to further bureaucracy. This issue must be discussed further.

Private and Local Actors

The requirement that only official government endorsed activities to increase adaptive capacity may be funded through the FCCC regime precludes private and local community initiatives from receiving funding. The COP has not addressed this issue in its decisions on the set up of the NAPA and the LDC Fund or through any of its guidance documents. Government endorsement is still a precondition, and, thus, a government's choice of priority activities may be influenced politically during the preparation of the NAPA. Especially for developing countries with weak political governance systems, the funding community should explore ways of enabling projects to enhance adaptive capacity without government approval or involvement. This could be achieved by entrusting private development agencies with the implementation of projects and programs rather than channeling funds through the GEF and its implementing agencies. On the other hand, again, such direct funding would create a problem of legitimacy in the recipient country; i.e., why would one NGO or community receive funding and not the other? Such issues should be discussed and addressed by developing country governments and civil society.

Stage III—Real Adaptation Funding

The Parties must agree on how to ensure funding for Stage III activities. Since the obligation to finance measures to implement Article 4.1(b) is included in the Convention, such activities cannot be restricted to the Adaptation Fund alone. Setting in place measures to enhance adaptive capacity in developing countries is especially relevant in this area. It remains unclear how the Convention funding mechanisms will cooperate with general ODA and other convention funding (for instance, through the Convention for Biological Diversity). Because the GEF has a strong consensus culture with both developing and industrialized country members, both the climate regime and the GEF could be a forum to develop a strategy that meets both recipient and donor country needs.

Endnotes

[1] UN Framework Convention on Climate Change, (1992) 31 ILM 31, 849; Kyoto Protocol to the UN Framework Convention on Climate Change (1998) 37 ILM 22, all available at www.unfccc.int.

[2] An Inter-Agency Task Force and Secretariat have been established as the main instruments for the implementation of the Strategy. The participation is optional for countries; see www.unisdr.org.

[3] The Declaration defines the right to development as (Article 1) "an inalienable human right by virtue of which every human person and all peoples are entitled to participate in, contribute to, and enjoy economic, social, cultural and political development . . ." Article 3 of this declaration was interpreted to give developing nations a right to official development aid, if not to a particular amount. The legally binding nature of this part of the declaration was rejected by industrial states and thus, not accepted as a principle of international law. See for more analysis Chowdhury et al. (1992) and Bulaijic (1993).

[4] Available at www.paris21.org. The seven goals, also contained in the UN Millennium Declaration, are halving the proportion of people living on less than $1 a day; enrolling all children in primary school; empowering women by eliminating gender disparities in education; reducing infant and child mortality rates; reducing maternal mortality ratios; promoting access to reproductive health services; and promoting environmentally sustainable development. In reference to the latter, the report states: "There should be a current national strategy for sustainable development, in the process of implementation, in every country by 2005, so as to ensure that the current trends in the loss of environmental resources are effectively reversed at both global and national levels by 2015."

[5] Annex I of the FCCC lists all industrialized/developed countries plus economies in transition in central and Eastern Europe as well as Russia and Ukraine.

[6] COP 2 brought into force a Memorandum of Understanding (MOU) between the COP and the Council of the GEF which sets out the legal framework of the cooperation between the bodies. See FCCC/CP/1996/15/Add.1; available at www.unfccc.int.

[7] See for more detail the GEF report on the development of a climate change strategy, contained in FCCC/CP/1995/4, the reports of the GEF to COP2 in FCCC/CP/1996/8, to COP4 in FCCC/CP/1998/12, to COP5 in FCCC/CP/1999/3, to COP6 in FCCC/CP/2000/3 and FCCC/CP/2000/3/Add.1 and to COP7 in FCCC/CP/2001/8; available at www.unfccc.int. See also Mertens (1994) and Ehrmann (1997).

[8] Donors contributed an overall sum of US$2.966 billion (see the Beijing Declaration of the second GEF Assembly, available at gefweb.org).

[9] See UNGA Resolution 49/122 endorsing the outcomes of the Global Conference on the Sustainable Development of Small Island Developing States (Bridgetown, Barbados) 26 April to 6 May 1994 (hereinafter The Barbados Programme of Action); and UNGA Resolution 50/11. For information on the follow-up program, see www.sidsnet.org. For the SOPAC project developing a global environmental vulnerability index, see www.sopac.org/Projects/Evi/index.htm.

[10] In the framework of the FCCC, such a prioritization has not been demanded by the group of developing countries (G77 and China)—possibly to avoid conflicts about funding eligibility within the group.

[11] Additional guidance to the GEF was provided at COP2 through Dec. 11/CP.2, FCCC/CP/1996/15/Add.1. at COP4 through Dec. 2/CP.4, FCCC/CP/1998/16/Add.1, and at COP6bis FCCC/CP/2001/L.4/Rev.1, which was reiterated at COP7, Decision 6/CP.7, FCCC/CP/2001/13/Add.1; see www.unfccc.int.

[12] To date, six projects have been approved, most of them regional in character. One of the Stage II projects is the joint UNEP and IPCC project: "Assessment of Impacts and Adaptation to climate change in Multiple Regions and Sectors". See GEF (2002) and, for background information, www.gefweb.org.

[13] All Annex II countries must report the status of implementation of the financial obligations in their national communications, see FCCC, Article 12.3. See also the article in the NGO newsletter Eco, "Looking Forward to the Funding" (9 November 2001), available at www.climatenetwork.org/eco/. For the DAC and OECD practices in this area, see www.oecd.org.

[14] Parties to the Montreal Protocol decide annually on budgets and contributions by Parties (Protocol on Substances that Deplete the Ozone Layer (hereinafter Montreal Protocol), (Montreal), 16 September 1987, 26 I.L.M. (1987), 154, Article 10.7). The contributions to the UN are assessed on the basis of GNP, conversion rates, and expected debt burden and adopted by the UN General Assembly. A new scale of assessment was adopted in December 2000, lowering the ceiling of the amount to be paid by any single country from 25 to 22 percent of the budget. See also U.N. Doc. A/55/521 and Add.1-3 and Resolution A/RES/55/5 B-F (www.un.org). Germany now pays 9.8 percent, Japan pays 19 percent, and the United States pays 22 percent of the overall UN budget.

Acknowledgments

I would like to thank Richard Klein (PIK) for his advice and support, Richard Tol (Professor for Global Change and Sustainability, University of Hamburg), Philipp Knill (German Ministry for Economic Cooperation and Development), and the German Scholarship Foundation (Studienstiftung des deutschen Volkes) for the financial support provided.

References

Bodansky, D. (1993), The UN Framework Convention on Climate Change: A Commentary. *Yale Journal International Law* **18**, 451.

Bulaijic, M. (1993), *Principles of International Development Law*. Oxford University Press.

Chowdhury, F. et al. (eds.) (1992), *The Right to Development in International Law*. Kluwer Law.

Ehrmann, M. (1997), Die Globale Umweltfazilität (GEF), 57 ZaÖRV (Zeitschrift für ausländisches Öffentliches Recht und Völkerrecht), 565 (with English summary).

GEF (2000), Review of Climate Change Enabling Activities. Evaluation Summary Report #2-00, November, available at gefweb.org.

GEF (2002), Report of the GEF to the Eighth Session of the Conference of the Parties to the UN FCCC. Contained in document FCCC/CP/2002/4, available at www.unfccc.int.

International Federation of Red Cross and Red Crescent (2001), *World Disaster Report 2001, Focus on Recovery*. Geneva.

Klein, R. (2001), Adaptation to Climate Change in German Official Development Assistance. Deutsche Gesellschaft für Technische Zusammenarbeit (GTZ), available at www.gtz.de.
Mertens, K. (1994), Towards accountability in the restructured Global Environment Facility. *RECIEL* **3**, 105.
Oberthür, S. and Ott, H. (1999), *The Kyoto Protocol.* Springer, Berlin.
Werksman, J. (1993), Incremental Costs under the Climate Change Convention: The Legal Context. FIELD (Foundation for International Environmental Law and Development) Working Paper.
Yamin, F. (1998), The Clean Development Mechanism and Adaptation, Paper for the FCCC Secretariat Capacity Building Workshop for Project Based Mechanisms, 17-18 September 1998. FIELD Working Paper No. 19.

10. Adaptative Capacity: The Philippines Coastal Resources Experience

Rosa T. Perez

Natural Disaster Reduction Branch

Philippine Atmospheric, Geophysical and Astronomical Services Administration (PAGASA)

Department of Science and Technology

Weather and Flood Forecasting Center, Quezon City, Philippines

Many of the world's coastal areas and small islands are highly vulnerable to climate change. Coral atolls, low-elevation reef islands, low-lying deltas and barrier coasts are especially sensitive to sea level rise and changes in rainfall, storm frequency and intensity. The potential impacts of sea level rise could have negative effects on tourism, water resources, fisheries, human settlements, financial services and human health. The Philippines, an archipelago with long and irregular coastlines that are vulnerable to many other natural hazards, is particularly vulnerable to sea level rise. Population growth is steadily increasing while valuable ecosystems are continually declining. Also, tidal gauge observations indicate already increasing sea levels attributable to high sediment loading and siltation, especially near Manila, the seat of the national government. Given the long history of storm surges and floods in the Philippines, sea level rise due to global warming will aggravate the flooding potentials for low-lying coastal areas. Several response strategies to current climate variability and nonclimate factors could also address problems that may arise because of future climate change.

1. Introduction

Current scientific knowledge indicates with increasing confidence that the unconstrained releases of greenhouse gases from most human activities will ultimately lead to global climate change (McCarthy et al., 2001). However, climate change models offer no specific conclusions about associated impacts at regional and local levels on the overall economy, because assumptions about future population and economic growth, technological changes and other factors largely differ.

In the Philippines, as in most developing countries, the issue of climate change seems to be buried under other, immediate concerns such as alleviation of poverty, economic security and environmental degradation, among others. Yet climate change is inexorably linked with economic activities crucial to modern societies—energy production and consumption, transportation, agriculture and forestry, real estate, coastal and marine resources use, industry and manufacturing and so on. As such, the issue cannot really be separated from the fundamental concerns of human society such as national economic planning, public administration and the quality of life for individuals, families and communities. At present, any responses to extreme climatic events are looked at from the point of view of disaster preparedness and mitigation.

The Philippines, as an archipelago with long and irregular coastlines that are often subject to natural hazards, is particularly vulnerable to the impacts of climate variability and change. Baseline information shows steadily increasing population growth and continually declining valuable ecosystems due to unsustainable usage. Also, tidal gauge observations indicate increasing sea level, especially for Metro Manila, the seat of the national government. The increased level has been attributed to high sediment loading and siltation. With climate change, specifically sea level rise, the vulnerability of the Philippine coastal environment is likely to be increased; this vulnerability will in turn intensify the impacts of unsustainable development.

Under dynamic climatic conditions, the suggested framework for adaptation of coastal resources is integrated coastal management, or ICM. ICM offers a window of opportunity for coastal managers to meet the present challenges of environmental degradation as well as climate variability. It is also hoped that such an approach will work with the long-term problem of climate change.

2. Vulnerability of the Philippine Coastal Sector to Climate Variability and Change

2.1 The Coastal Environment

In the Philippines, about 80 percent of the provinces and 65 percent of cities and municipalities share the coast. The coastal zone supports major economic activities such as farming, fishing and tourism, and almost 70 percent of the total population, about 72 million in 2000 (NEDA, 2001), resides near or in the coastal areas. Most are "informal communities", economically depressed with little or no sanitation facilities or social support systems. We can see in Figure 1 an example of a high-density populated area in Metropolitan Manila. During the rainy season and at times of high tides, such areas experience frequent flooding, as shown in Figures 2-4.

During extreme warm periods associated with the El Niño phenomenon, toxic algal blooms cause fish kills, affecting the livelihood of fishing communities that rely on the daily catch from municipal waters (waters within 15 kilometers of the shore) for subsistence (see Box 1). This adds to existing environmental and socioeconomic problems in the coastal areas. These problems include declining fisheries productivity due to overharvesting and loss of habitat, increasing environmental damage, rapid population growth and migration to coastal areas and widespread poverty among the coastal communities.

Figure 1. A high density populated area in the coastal area of the Philippines. Most of these are informal communities and belong to the economically depressed areas.

Figure 2. Frequent flooding results when heavy rains coincide with high tide. People's lives are disrupted because they have to move their property or evacuate their flooded homes.

Figure 3. Work and classes are suspended during floods, which are usually chest deep.

Figure 4. Even a vehicle for emergency use cannot avoid being stalled in a flooded street, a common sight during heavy rains and high tide occurrence.

Box 1. A recent newspaper article on the effects of a red tide on the Philippine coast

The Bureau of Fisheries and Aquatic Resources (BFAR) is testing water and dead fish samples after hundreds of coral reef fishes were found lining the seven-kilometer stretch of coastline in the Balayan Bay, a known premier eco-tourism spot in the country, over the weekend.

BFAR Director Malcolm Sarmiento told **The STAR** that an initial investigation revealed that a mild red tide caused the fish kill that was discovered by local folks late Friday initially along the two-kilometer stretch of coastline near Lian, Lemery, and Anilao in Batangas.

However, Sarmiento said the initial findings could not certain if the contamination originated from the waters of the bay, as the mild red tide could have been from other nearby bodies of water or areas.

"We are still pursuing a deeper examination on the water samples and samples of dead fishes," he noted. "It is possible that the affected area was actually from somewhere else, that the dead fishes were just brought in Balayan because of the strong wind or tide."

During an earlier monitoring, the BFAR, Philippine Coast Guard, Coral Cay, Bantay Dagat and World Wide Fund-Philippines (WWF) found that there was an abnormally high water temperatures of 33 degrees Celsius and a profusion of single-cell algae, which concentration appeared to be in Brgy. Locloc, Bauan.

Patches of red film on the water surface were also noted.

Since the Balayan Bay is a popular tourism destination, the public was also warned against swimming in the affected waters to avoid the possibility of skin irritations until the area is declared clear from contamination.

Katherine Adraneda, Arnell Ozaeta, The Philippine STAR, June 11, 2002

The most valuable coastal ecosystems in the Philippines are the mangroves and the coral reefs. Mangrove forests are a unique feature of protected coastal shorelines of the tropics and subtropics; their root systems stabilize sediment, dampen wave energy, provide habitat shelter for numerous organisms and provide the basis for the near-shore marine food web. The best-developed mangrove forests are in areas of high rainfall and stream runoff. Thus future changes in rainfall patterns and runoff will have severe impacts on mangroves. Mangroves grow best in moderately saline environments and could keep up with sea level rise of up to 12 centimeters per 100 years. However, the current rate of mangrove loss to overcutting, land clearing and habitat conversion suggests that the effects of climate change may be a minor factor in the fate of this important habitat. The mangroves of the Philippines have steadily decreased in quantity and quality over the years: the current area is about 100,000 hectares, compared to 450,000 hectares at the beginning of the twentieth century (DENR-EMB, 1996).

The Philippine coral reefs cover about 27,000 square kilometers and are about 15 to 30 kilometers deep. They contribute from 10 to 15 percent of the total annual fish yield, equivalent to between 5 and 24 metric tons per square kilometer of economically valuable fish and invertebrates (DENR et al., 1997). Aside from fish recruitment downstream of the corals, the reefs offer biodiversity benefits, which are increasingly recognized as important in medicine, research, education and tourism, among others. Reefs also provide natural protection to shoreline structures against wave and storms by acting as an effective barrier to erosion in low-lying areas. Economic returns derived in tourism activities include fees from scuba diving, snorkeling and other recreational activities.

However, in terms of live coral cover, the overall condition of the coral reefs in the country is not good. Only about 5 percent of the Philippine corals are in excellent condition, attributed to serious anthropogenic stresses that include sedimentation, pollution and physical alteration of coral reefs (DENR et al., 1997). Sedimentation is usually caused by poor land use practices, including deforestation. Pollution and overnitrification come from domestic agricultural and industrial wastes. The reefs are physically altered by coastal construction projects, destructive fishing practices (e.g., poison and blast fishing), ship grounding and coastal tourism, which bring millions of eager divers and snorkelers to coral-rich areas. Scientists predict that within the next 20 years, increasingly more coral reefs will succumb as the intensity and frequency of the anthropogenic stresses increase along with the size and density of coastal human populations, as well as the impacts of the changing climate (ADB, 2001). Corals are greatly affected by changes in sea surface temperature (viz. the 1983, 1987 and 1997 bleaching events during El Niño episodes), but the impact of sea level rise is also expected to be important. Higher sea levels may decrease the amount of light reaching the corals at the sea bottom, affecting photosynthesis of the tiny plants providing foods for the coral polyps. Destruction of coral reefs translates to economic losses for the country. Various estimates in the Philippine coral reefs showed that a healthy reef could produce sustainable catch of 200 tons of reef fish in 10 years, while a destroyed but recovering reef could produce only 72 tons for the same period (White, 1989). Based on the year 2000 price index, this was approximated at US$192,000 (about 9.6 million pesos).

2.2 Current Climate Risk

The country is highly susceptible to environmental damages and prone to natural disasters because of its location, mountainous terrain, fragile soils, low-lying coastal areas and largely variable climate (specifically rainfall). Extreme weather and climate events (e.g., drought, strong tropical cyclones, monsoon rains) affect almost all major sectors, including agriculture, water resources, coastal resources, human settlements and human health. The coast is frequently subject to storm surges and riverine flooding from tropical cyclones. An average of 20 tropical cyclones pass yearly through the Philippine area and on average about nine of them cross land (CAB, 2000). Flooding usually occurs in the low-lying coastal areas during heavy rains. This usually aggravates the already heavy soil erosion condition and triggers landslides in some areas (Figures 5 and 6). The eastern side of the country receives heavy rains during the northeast monsoon season (November to February), while the western side receives them during the southwest monsoon season (from May to September). The southwest monsoon season coincides with the typhoon season as well. Tropical cyclones cause heavy damages, which seriously affects the economy of the nation. For example, a typhoon hit Central Visayas in 1990 and caused damages worth P10.85 billion (US$43.4 million), while another typhoon hit Southern Luzon in 1995 with damages of P9.30 billion (US$33.2 million) (NDRB, 1999). Year-to-year climate variability greatly affects the amount of rainfall in the country, with El Niño causing drought (below average rainfall) and La Niña causing floods (above average rainfall).

2.3 Future Climate Risk

Increasing coastal population and development has already resulted in scarce and limited space and resources, creating use conflicts as well as socioeconomic instability. With climate change and sea level rise in the picture, the vulnerability of the Philippine coastal environment is likely to be increased, which in turn will intensify the impacts of unsustainable development.

Considering the long history of storm surges and floods in the Philippines, sea level rise will aggravate the flooding potentials for low-lying

Figure 5. Soil erosion is heavy in some coastal areas, which are usually aggravated by heavy rains and high energy waves

Figure 6. Landslides are triggered by heavy rains especially on very unstable slopes

coastal areas. A 1-meter sea level rise poses a potential threat of inundation to about 2,000 square kilometers of low-lying areas, islands and islets (NAMRIA, 1992), placing more than two million Filipinos in danger. The coastal resources sector is also expected to be exposed to more frequent and more intense tropical cyclones.

Because of existing stresses to the natural system, the vulnerability of the Philippine coastal resources and infrastructure (Figure 7) to sea level rise

Figure 7. The sea is moving inland. Some infrastructures built near the coast 10 or 15 years ago are now always flooded with seawater.

increases constantly (Perez et al., 1996), underscoring the need for an integrated framework to address these issues. The inadequacy of data and lack of either suitable information systems or a coordinated institutional structure in managing coastal and marine resources compound the difficulties.

3. Capacity to Respond to Climate Variability and Change

Under the United Nations Framework Convention on Climate Change (UNFCCC), the Philippines is committed to prepare a national action plan to address vulnerability and adaptation to climate change impacts in the context of sustainability. For the coastal resources sector, there are two ways to address these issues. One is to strengthen the sector's development programs to enhance the sustainable use of coastal resources; and the second is to decrease vulnerability to the potential impacts of climate change.

An analysis of coastal vulnerability starts with some notions of the natural system's potential to be affected by the different biogeophysical effects of sea level rise, rainfall and wave pattern changes, and of its natural capacity to cope with these effects. Ideally, a coastal vulnerability assessment would incorporate a detailed quantification of the predicted impacts, based on an extensive knowledge of their space and time dynamics. However, the unavailability of basic data on coastal topography and geomorphology pose difficulty in determining accurately the impact zones. Likewise, the potential impacts of climate change in the coastal zone could include the danger of worsening of some current coastal problems as well as the emergence of new ones. Hence, assessing the potential impacts of climate change can begin with a focus on the current impacts observed within the context of the present-day climate variability. If one can account for the effects of the present-day climate and responses to current vulnerability, one will be in a better position to deal with future changes. Also, any attempts to reduce vulnerability to current climate can bring immediate or near-term benefits, which are unlikely to be regretted.

Other methods consist of semi-empirical assessment, review of past work, limited coastal area ground surveys, data collection from different governmental agencies on coastal resources and their management, and focus group meetings with government agencies and nongovernmental organizations (NGOs). Based on the data required and collected, trends are established to get appropriate indicators of vulnerability.

Adapting to sea level rise and to other effects of climate change involves balancing environmental, economic, social and cultural values. How coastal resources will be affected will primarily depend on the local patterns and intensity of climate change and also on the local coastal environment. Several response strategies have been identified to address the vulnerability of the Philippine coastal resources to present climate variability, nonclimate factors and, to some extent, future climate change. Box 2 presents the results of an adaptation planning process at the local (provincial) government level.

Box 2. Addressing climate variability and change in a local planning workshop

Between April and November 1999, eight Philippine provinces participated in a local planning workshop to consider the effects of climate variability and change (papers were presented at the workshop, entitled the Local Action Planning for Climate Change Impacts in Selected Provinces; this was a cooperative effort with the Foundation for Philippine Environment and the Institute of Climate, Energy and Environment). Most of the issues, listed below, are related to current problems in the use of coastal resources which result in the dwindling, degradation or destruction of coastal resources and environment. Salt-water intrusion was also pinpointed as a growing problem. The occurrence of coastal hazards, specifically flooding and erosion, was identified as one of their concerns, as was potential impacts of climate change on coastal resources.

Box 2. Addressing climate variability and change in a local planning workshop (cont.)

Problems in participating coastal provinces related to sea level rise, climate variability or climate change

Directly caused or affected by sea level rise, climate variability or climate change	• Salt water intrusion • Displacement and dislocation of communities due to flooding and other hazards • Submergence of small islets • Inadequate supply of potable water during summer months • Flooding • Lack of potable water sources/supply • Susceptibility to natural hazards
Aggravate the impacts of or aggravated by sea level rise, climate variability or climate change	• Coastal water pollution • Coastal erosion, siltation and sedimentation • Sand and gravel quarrying and coral poaching • Loss of biodiversity • Coastal/marine resources destruction • Unregulated development along the coastal zone • Reduced fish catch and aquaculture production • Illegal fishing • Destruction of coral reefs/dwindling mangrove resources

Generally, most participants had little awareness of climate and climate change concepts. However, they were highly knowledgeable of developmental activities in their province, and especially of their effects on the protection and enhancement of the coastal environment. The workshop activities gave the participants an opportunity to effectively integrate strategies for evaluating the vulnerability of provincial coastal resources and adapting to the effects of climate change with strategies for managing other coastal concerns and problems, e.g., coastal pollution and habitat protection.

The coastal zones of the participating provinces are already experiencing multiple problems. Among these are declining ecosystems (coral reefs, sea grasses, mangroves); low yield fish stocks; fast growing populations; human settlements (squatting); and conflicts in use of the coastal zones and resources, among others.

Fishing is among the most important economic activities in the Philippines, generating P 81.2 billion in the previous years (DENR et al., 1997). However, fish yields are consistently declining. One reason is the conversion of mangrove areas to fishponds. This increases the output of aquaculture but lowers the rate

Box 2. Addressing climate variability and change in a local planning workshop (cont.)

for replenishment of coastal fish stocks. Most coastal dwellers depend on fishing as the main source of income, and yet they continue to believe that the supply of resources from the sea is inexhaustible and that the coastal ecosystems can continue to withstand any human interference, including climate change impacts.

One of the most important coastal ecosystems, mangroves, are a unique feature of a protected coastline. Their root systems can stabilize sediment and dampen wave energy, thus preventing coastal erosion. Hence, the depletion or destruction of this valuable ecosystem will render the coastal area highly vulnerable to climate change impacts such as sea level rise, changes in water currents, more intense tropical cyclones and higher storm surges. However, most of the fisher folks in the participating provinces are understandably more concerned about day-to-day survival than about the future.

At the workshop, the participants identified several strategies, either to enhance the capacity to adapt to future climate change or as direct measures to address present and future climate risks (see list below). Most of the prescribed responses to the identified issues are nonstructural. That is, they invoke the creation, enhancement or strict implementation of coastal laws (e.g., RA8550, mining laws), regulations (e.g., on land use, zoning) or programs (e.g., Disaster Management Program, Coastal Environment Program, Coastal Zone Management). One province exceptionally identified the use of an indigenous plant, the *ipomea*, as a least cost method to stabilize the coast in place of a more expensive engineering solution to coastal erosion. Information campaigns and advocacy were also popular, to increase the awareness of the populace, which in turn would increase their capacity to adapt to future climate change. Some provinces realized that monitoring sea level rise and other indicators of climate changes, combined with data obtained using geographical information systems (GIS), is also an important strategy for decision support. Mapping vulnerability to sea level rise was also seen as a critical activity.

Current/proposed measures to enhance capacity to adapt to climate variability and change

Adaptation measures	• Modify setback policies to address climate change/sea level rise • Conduct research studies on salt water intrusion, fisheries and aquaculture • Strengthen Disaster Management Program • Improve typhoon warning system • Institute flood prevention/protection measures • Stabilize shorelines/prepare maps of hazards and vulnerability to floods and sea level rise • Stop further conversion of mangrove into fishpond development

Box 2. Addressing climate variability and change in a local planning workshop (cont.)

Current/proposed measures to enhance capacity to adapt to climate variability and change (cont.)

Adaptation measures (cont.)	• Put in place ICM and expand Coastal Environment Program (CEP) • Undertake massive upland and coastal reforestation, including expanding community-based mangrove reforestation program • Implement information, education and communication (IEC) awareness program as part of decision support mechanism: • Monitor sea level rise and climatological data: tidal gauge stations (costly) vs. indigenous methods (staff gauges) • Install GIS
Capacity enhancing measures	• Empower people in the management of coastal resources • Inventory and survey coastal resources • Institute provincial environmental and natural resource accounting • Require industries to install desalination facilities • Expand coverage for artificial reefs, marine sanctuary, marine reserves • Strengthen coordination between Department of Environment and Natural Resources (DENR) and LGUs • Enforce appropriate land use and zoning • Institute strict monitoring and enforcement of mining laws (sand and corals) and other coastal management policies, laws and regulations • Formulate comprehensive coastal development plan • Develop/improve watershed management, including identifying and developing potable water sources • Reactivate/reorient Environment and Natural Resources Committee (ENRC) in the coastal municipalities • Implement *Lingap para sa Mahihirap* (Poverty Alleviation Program) • Regulate installation of water pumping systems • Strengthen/enhance integrated waste management program, including adopting coastal clean-up movements • Provide alternative livelihood and resettlement program

Box 2. Addressing climate variability and change in a local planning workshop (cont.)

Some of the adaptation measures being implemented are specifically to strengthen the coastal ecosystem. These are currently carried out on a pilot basis within a limited time period, especially through foreign-funded projects. There are cases where financial support for a particular program has been depleted, and some measures are left unfinished or abandoned (e.g., mangrove reforestation and coral rehabilitation). Other identified barriers to the successful implementation of adaptation measures include the lack of political will of LGUs, lack of technical knowledge at the local level and, to a certain extent, the negative attitudes or reluctance of people to change. It was also suggested that it might be useful to identify which adaptations make sense without climate change or accelerated sea level rise and help with climate change or accelerated sea level rise.

The present management strategy in the Philippines combines the goal of sustainable development and the integration and interdependence of issues and sectors among and between environment and economic development. New practices have pointed out that environmental protection and economic development cannot be considered separately. ICM provides an opportunity to carry out a continuous, iterative and consensus building process to achieve a set of goals, including adapting to the effects of climate change. ICM is seen to be effective, since its principles are based on sustainable development.

In the Philippine context, the key requirements for sustainability, as they apply to fisheries and other coastal-based industries, tourism, coastal development and protection of sensitive or critical areas, include the following:

The precautionary principle approach is used in the management of coastal resources.

Decisions about planning and management are based on the best available information. Incomplete information should not be used to postpone precautionary measures intended to prevent serious or irreversible harm to coastal resources. The higher the risk of unacceptable levels of change or of serious environmental impacts, the more conservative the measures required to reduce or remove the risk.

Coastal stakeholders are given opportunities to actively participate in all aspects of coastal management.

Coastal resources are considered to be mainly in the public domain. Hence public and major stakeholders must have a central role in planning, implementing, operating, monitoring, enforcing and evaluating ICM plans and programs. It is the responsibility of the government to ensure that the appropriate and effective opportunities for such public participation are available. Table 1 lists the major stakeholders and the roles they could assume in integrated coastal management. Among the many advantages of involving the stakeholders in coastal management is the sense of ownership or stewardship developed in the members of the community toward the protection of coastal resources. Some of the most successful mangrove reforestation and coral rehabilitation work in the Southern Philippines is community based.

Table 1. Coastal resources sector stakeholders and their roles in the integrated coastal management

Stakeholders	Roles
National government	• Manages coastal areas and resources for the benefit of the people • Performs highly specialized functions such as impact assessments, research, science and technology, economic development • Provides funds
LGUs	• Monitor the implementation of coastal rules and regulations in the localities • Coordinate with the national government for smooth operational functions among LGUs who have jurisdiction in one coastal zone area.
Coastal/ocean users (e.g., fisheries, tourism development and recreation, aqua culture, military, shipping and port operations, mining, subsistence activities and offshore oil operations)	• Articulate their special needs and concerns regarding coastal ocean space and resources • Ensure stewardship of coastal ocean space and resources

Table 1. Coastal resources sector stakeholders and their roles in the integrated coastal management (cont.)

Stakeholders	Roles
NGOs and POs (e.g., women's organizations, religious groups, local and international environmental NGOs; people's organizations and youth groups)	• Organize the community • Work with the citizens to assess the main priorities and needs • Conduct community education programs • Provide feed backs to government agencies • Monitor and manage resources • Act as public advocates • Provide extensive knowledge about the quality and quantity of coastal resources and their use
Landowners	• Ensure stewardship of land and special habitats • Avoid erosion, flooding and disasters
Businesses (in or near coastal areas)	• Provide capital for development projects, facilities and equipment • Provide investments that lead to employment opportunities • Pay taxes that can help finance ICM projects.
Users of coastal and upland resources (agriculturists, foresters, loggers and miners)	• Adopt sustainable, environmentally benign, soil management and water allocation practices.
Universities, scientific institutions and other educational entities	• Raise public awareness through outreach activities • Provide data and information for making informed management decisions • Develop special education and training programs • Serve as advocates for rational, scientifically based management of oceans and coastal resources.

Responsibility and resources must be fully devolved in local level decision-makers, who are closest to the resources to be developed.

Local government units (LGUs) and coastal communities are closest to the problems and solutions in coastal management. Devolution of responsibilities for the coastal zone and municipal waters to the lowest tiers of local government is essential to coastal resource rehabilitation efforts. Adequate financial resources must be invested in coastal management to sustain benefits derived from coastal resources.

The maintenance of healthy and productive ecosystems is fundamental to the management of coastal areas.

Managing activities that affect coastal resources requires improvements in the understanding of living and nonliving resources and processes. Integration between and among sectors, institutions, geography—and disciplines, and social, economic, cultural and environmental aspirations—should be assessed and accommodated through integrated planning and management of the multiple uses of coastal resources.

Coastal resources should be allocated to uses that provide the greatest long-term community benefits that are compatible with the maintenance of the ecosystem's health.

If the potential impacts of an action are of concern, the first priority should be the maintenance of a healthy coastal ecosystem. Short-term economic gain for a few should never be the basis of a decision that has adverse environmental impacts in the long term.

4. Opportunities to Enhance Adaptive Capacity

4.1 The Local Government Code and ICM

Under the 1991 Philippine Local Government Code (LGC), the primary mandate for managing coastal resources in municipal waters, which extend seaward to 15 kilometers from the shoreline, was devolved to 832 coastal municipalities (55 percent of municipalities nationwide) and 57 coastal cities (68 percent of cities nationwide). The capacity of the LGUs to assume this mandate has evolved over time. Many basic innovations were needed over 10 years to achieve the level of governance now provided by the LGC. It was not until 1998, however, that LGUs become more aware of and concerned with their responsibilities in environment management, including, to a lesser extent, the concerns of climate change. By 1999, decentralization was almost complete, with national government technical assistance increasingly more consistent with the devolution process. Local government and communities introduced innovative mechanisms for effective resource management, using intergovernmental agreements and public-private partnerships to comanage resources across jurisdictions. Coastal resource management is being articulated as a basic service function of the LGUs.

The precautionary principle of ICM provides a favorable entry point for addressing climate change related concerns. Impacts in the coastal zone (such as increased erosion, flooding and saltwater intrusion) are best managed within the framework of an ICM program. Effective strategies for adapting to climate change integrate ICM with the other major building blocks of the National Climate Change Action Plans. In the Philippines, guidelines and legislation for the implementation of an integrated coastal management program for all coastal zones in the Philippines and the Land Use Policy Bills are currently under review in the Legislative Branch. Appropriate land use planning in coastal zones is envisioned to reduce vulnerability to a rise in sea level, through the inclusion of setback mechanisms, allocation of low-lying vulnerable land to lower value use such as parks rather than housing and requirement of compliance with construction standards or post-storm reconstruction standards. These policies reduce the risk of living in coastal areas from current climatic variability and protect against future sea level rise. Public easements and buffer strips are proposed to be treated as separate lots during land surveys and, as such, should be excluded from tilling or private ownership. In the integrated coastal zone management Policy Bill, the LGUs are required to reserve foreshore areas that are critical areas for recreation and tourism and other public uses and are excluded from disposition. Inclusion of wetlands, swamps and marshes in the National Protected Areas (NIPAS) with a category of wildlife sanctuary or unique ecosystem may be necessary. Preservation efforts should be made to maintain wetlands, swamps and marshes that are vulnerable. Institutionalized resource development such as those of mangroves and corals are proposed through the formulation of additional policies and regulations or amended existing policies and regulations. This is hoped to allow effective and sustainable resource management, for example, highlighting the massive reforestation of degraded mangrove systems through a community-based approach.

4.2 Environmental Disaster Management

Under the integrated coastal management strategy, a multihazard mitigation or protection plan for natural coastal hazards could be developed, with priority on the maximum reduction of threats to life, structures and economic production. New anticipatory approaches will be needed to increase the resilience of vulnerable areas to and improve their recovery from future disasters, including that of the probable impacts of climate change and sea level rise. Under this guidance, mining laws and reforestation of denuded watersheds should be strictly implemented to reduce river and coastal erosion. Geological, hydrometeorological and structural engineering evaluation is now required as part of the environmental impact assessment process prior to coastal development.

New opportunities could be explored, such as stopping government subsidies or tax incentives that support development of land sensitive to sea level rise like barrier islands, coastal wetlands, estuarine, shorelines and critical wildlife habitats. On the contrary, insurance and banking industries could be encouraged to factor risks of climatic variability into investment decisions, thereby reducing reliance on government-subsidized insurance and disaster relief.

These are only among the more general strategies to address future impacts of sea level rise and other climate change impacts in the coastal resources sector. Detailed actions should be site specific and must be done after a thorough impact analysis of the area involved.

4.3 More Options and Opportunities

While working on the framework of sustainability, ICM offers tangible incentives in the short term with easily recognizable returns to the participating community. Among these are:

- food security and other economic opportunities
- improvement of fish catch/production
- Coastal Resource Management Certification (CRMC) for the LGU
- good institutional structure
- provision of alternative livelihoods
- intergenerational equity
- ecological/biodiversity integrity, resource generation
- financial assistance
- clearly defined and expanded local roles in ICM
- empowerment of communities, increased sense of ownership
- improved coastal resource base
- political support
- provision of training, technological transfer assistance.

5. Challenges and Needs

As indicated in the local planning workshops, several barriers and needs must be addressed to be able to adapt or to increase the capacity to adapt to the changing climate.

5.1 Barriers

Several barriers to fulfilling the mandate of the LGC by local officials have been noted. These include, among others:

- lack of trained staff or expertise on coastal management
- low participation level and ecological literacy of the community
- unclear legal jurisdiction of local officials over resources
- low interest of local officials in coastal resource valuation
- inadequate funding
- lack of political will and leadership
- global, national and local market forces—demand for corals and aquarium fishes in foreign markets has been so great that even local officials have been involved in smuggling them out of the country.

5.2 Research and Training Needs

Improving the Scientific and Information Base for ICM

Any coastal management contains uncertainties about various aspects of the coastal systems. These include future trends of climate change and sea level rise, inadequate land use information and how technology affects pollutant discharges and habitat changes. Under the setting of LGC and ICM, integral components are data collection, assessment and directed research programs that develop inputs for the next round of analysis and planning information to reduce the uncertainties. Reliable data are required for the entire ICM process to:

- understand and analyze the present state of the coastal environment
- prepare the ICM program and evaluate its performance
- understand the resilience and vulnerability of the coastal system to climate change and sea level rise
- formulate effective response strategies.

Sea level rise (relative or absolute) is not the only effect of climate change in coastal areas. Climate change studies should combine the effects of sea level rise, changes in storm intensity and frequency, atmospheric circulation change and changes in rainfall patterns. Coastal management takes place in a dynamic setting and hence must be adaptive, i.e., responsive to the changing environment and new information. Being responsive requires a monitoring system to obtain information over time on variables that affect the choice of actions in the plan. Sea level, rainfall temperature, wind field, storm frequency, storm intensity and land use changes are examples of such variables.

Economic Valuation of Coastal Resources

The Philippine coastal resources have been undervalued historically, and are rarely factored into the national or local economic decisions. Continuous efforts must be emphasized to be conversant with new concepts and tools designed to provide quantitative measures of the environmental benefits of natural resources. These can be done through the economic valuation of natural resources (White and Cruz-Trinidad, 1998).

5.3 Education, Training and Outreach

Education, training and outreach activities that target all sectors of society are essential components in the successful implementation of ICM and climate change action plans. Universities, training and research organizations should develop and strengthen programs of research, education, and training, extension services, and technical assistance that will contribute to continuing ICM programs. These programs should combine theory and practice and must emphasize the applications of research to solve coastal management problems. ICM programs require teams with skills in resource and socioeconomics, ecology, geomorphology, coastal engineering, analysis of industrial and agricultural processes, financing and institutional (including legal) analyses. Addressing climate change also requires skills in meteorology, physical oceanography, earth science, geography and predictive computer modeling. The most difficult skill is integrating the various aspects of analysis and defining priorities among them, and discerning the long-range implications of current actions.

Training and education for ICM should be multidisciplinary, so trainees can become familiar with using all the scientific information related to ICM.

Training institutions should be encouraged to offer programs that satisfy the immediate needs of practitioners. The target groups should include elected officials; religious leaders; national, regional, and local policy officials; user groups; the private sector; nongovernmental organizations; and educators. The priority should include the development of courses to train the trainers.

Training is a continuous process and is preferably country-driven. Methods of training should be modified according to need and can take such forms as short courses, workshops, seminars, case studies, attachments, and internships.

Evaluation of training and appropriate career advancement should be used to ensure retention of trained personnel. Feedback mechanisms should be built in to update training programs.

Technology transfer is an important element in international environmental treaties and conventions. Equitable access to technology should be promoted through training and provision of equipment.

Climate change impacts on coastal zone systems are not well understood by the public or decision-makers. Hence, promotion of awareness about erosion, sea level rise, flooding risks and storm and earthquake standard building codes could be a cost-effective means of reducing future expenditures.

5.4 Financial Mechanisms

Financial resources are required to implement such government responsibilities as developing ICM action programs; conducting research, monitoring and evaluation; providing technical assistance; and increasing awareness through education and outreach programs. Public-private sector partnerships may be encouraged to support all these ICM objectives and activities. Among others, appropriations from government funds, revolving loan funds from international institutions, dedicated funds such as those derived from specific tax revenues and donor funds (international or bilateral donor agencies) may be possible sources for initiating and implementing any ICM, climate change programs or related projects. Based on recent climate change studies, impacts on coastal areas (such as increased erosion, flooding, and salt-water intrusion) are best dealt within the framework of an ICM program. ICM should be regarded as one of the major building blocks of the national climate change action plans, and should be implemented together with the other essential components to achieve integration and effective adaptation to climate change. Donor-assisted projects should include requirements for providing training to host-country counterparts.

6. Concluding Remarks

The variety of problems facing the coastal zones and the management of their resources necessitates the use of integrated management and community participatory approaches. Enhancing the capacity to adapt to the probable impacts of future climate change can be a part of the framework of an integrated coastal resource management. In the Philippines, ICM provides an opportunity to address the present problems in the coastal areas, which may not be climatic in nature, such as poverty alleviation and environmental rehabilitation. Currently, ICM planning is in the very short term and focuses more on immediate issues such as illegal fishing, blatant pollution and use conflicts, among others. There are, however, many entry points whereby policies can be promoted to support long-term interests, including climate change concerns. The formulation and enforcement of laws that could improve or rehabilitate coastal resources, such as those involving fisheries management, mangrove reforestation, coral rehabilitation, protection of shoreline and critical areas and coastal zone development plans, among others, follows the precautionary principle. ICM supports any measures that can strengthen the present system such that it could be robust enough to adapt to any changes in the future, including that of impacts from climate change. The Philippine example shows that whenever the community is involved, a sense of ownership is enhanced to protect and develop the coastal resources, so that its sustainable use is maintained. Decentralization of government helps in the successful implementation of ICM by providing more support to the local government, which is closest to the resources concerned. This is not to be misconstrued to mean that the local government will have the final say in the management of the local coastal zones. The LGUs still depend on the national government for the disbursement of funds and fielding of experts, and significant coordination and synchronization would be required between them.

References

ADB (2001), *Asian Environment Outlook: 2001*. Asian Development Bank, Manila, Philippines.

CAB (2000), PAGASA Tropical Cyclone Statistics, Information Brochure. Climatology and Agrometeorology Branch. Philippine Atmospheric, Geophysical and Astronomical Services Administration, Manila, Philippines.

DENR, DILG, DA-BFAR, and CRMP (1997), *Legal and Jurisdictional Guidebook for Coastal Resources Management in the Philippines*. Department of Environment and Natural Resources, Department of Interior and Local Government, Department of Agriculture—Bureau of Fisheries and Aquatic Resources, the Coastal Resources Management Project, Manila, Philippines.

DENR-Environmental Management Bureau (1996), Philippine Environmental Quality Report (1990-1995). Department of Environment and Natural Resources, Manila, Philippines.

McCarthy, J. J., Canziani, O. F., Leary, N. A., Dokken, D. J. and White, K. S. (2001), *Climate Change 2001: Impacts, Adaptation, and Vulnerability. Contribution of Working Group II to the Third Assessment Report of the Intergovernmental Panel on Climate Change*. Cambridge University Press, Cambridge, UK.

NAMRIA (1992), *1992 Report*. National Mapping and Resources Inventory Authority, Manila, Philippines.

NDRB (1999), Information Materials on Natural Disasters. Natural Disaster Reduction Branch, PAGASA, Quezon City, Philippines.

NEDA (2001), *Philippine Census Year 2000*. National Economic Development Authority, Manila, Philippines.

Perez, R. T., Feir, R. B., Gonzalez, E. and Carandang, E. (1996), Potential impacts of the sea level rise on the coastal resources of Manila Bay: a preliminary vulnerability assessment. *Water, Air and Soil Pollution Journal* **92**, 117-137.

White, A. T. (1989), Two community-based marine reserves: lessons for coastal management. In *Coastal Areas Management in Southeast Asia: Policies, Management Strategies and Case Studies*, Chua T. E. and Pauly, D., eds. ICLARM Conference Proceedings, **19**, 85-96.

White, A. T. and Cruz-Trinidad, A. (1998), *The Values of Philippine Coastal Resources: Why Protection and Management are Critical*. CRMP Document No. 37-CRM/1998. Coastal Resources Management Project, Manila, Philippines.

11. Adaptation to Climatic Variability and Change in Tlaxcala, Mexico

Cecilia Conde
Centro de Ciencias de la Atmósfera
Universidad Nacional Autónoma de México
México, D.F. México
Hallie Eakin
Department of Geography and Regional Development
University of Arizona, Tucson, Arizona, USA

This paper discusses anticipated impacts of climatic change and variability on the agriculture sector in the state of Tlaxcala, Mexico, and the potential of farmers to mitigate their vulnerability through improved crop management. The project involved close collaboration with farmers and researchers in Tlaxcala to identify and analyze climatic risk and impacts, and to explore adaptation options. The initial optimism for maize yields in Tlaxcala based on climate change scenarios introduced in a crop simulation model was tempered by the observed negative implications of the El Niño-Southern Oscillation for rainfall distribution and frequency of frost events in the state. Facilitating "appropriate" crop choice has been the primary focus of adaptation in the state, and there is some evidence of successful risk mitigation using seasonal forecasts in crop decisions. Further investigation revealed the importance of the socioeconomic context of farmers' decision-making and the need to consider household risks and opportunities in the promotion of adaptation options.

1. Introduction

Recent research in Mexico on the vulnerability of agriculture and water resources to climatic variability and change has illustrated the importance of flexibility in decision-making as a key element in adaptation capacity

(Conde et al., 2000; Eakin, 2000). Climate change scenarios and analyses of the effects of the El Niño-Southern Oscillation (ENSO) in Mexico have shown that agricultural yields, particularly in the central highlands, are particularly sensitive to changes in rainfall and temperature patterns. Populations dependent on rainfed agriculture, as well as populations located in regions where water is already a scarce and contested resource, will perhaps be the most affected by anticipated climatic changes. Under these circumstances, the study of possible climatic change conditions and the forecasts of ENSO events could be important tools in the design of coping strategies.

The capacity of such populations to adapt, however, depends in large part on the diversity of choices available to them and the flexibility of their livelihoods. This flexibility depends in part on political, economic and institutional factors and the influence of these factors in risk management. Furthermore, the empirical knowledge of environmental change held by vulnerable populations is often ignored as a resource in adaptation. This knowledge and experience not only can serve to improve climate impact scenarios and guide scientific research, particularly where observational data are absent, but also can serve as fundamental ingredients in the development of public policies that facilitate innovative adaptation to environmental change.

These issues are explored in a case study of adaptation in Mexico. We discuss the anticipated impacts of climatic change and variability on the agriculture sector in Tlaxcala in relation to the implications of economic globalization processes for vulnerable populations, illustrated in the region by the North American Free Trade Agreement (NAFTA), urbanization and social inequality. Collaboration with farmers in Tlaxcala on climate forecasting for agriculture has illustrated how adaptation capacity is closely tied to the broader socioeconomic context of decision-making. The empirical climate knowledge held by Tlaxcalan farmers has been an important tool in improving climate impact analyses for the state. In addition, their knowledge of climate risk management in production offers some possible avenues for improving the capacity for adaptation in the sector, if such strategies could be made compatible with broader economic and social policy.

Over the last decade, water availability has become a major concern for Mexico. The Ministry of Environment (SEMARNAT, 2001) has launched a campaign ("Crusade") for forest and water conservation in the country, since these issues are considered a matter of national security, particularly in the center and northern regions of the country, where population growth rates suggest that population could double by 2025. More than 80 percent of Mexico's territory is classified as arid or semi-arid (Reyes Castañeda, 1981). Water resources are not equally distributed. Seven percent of the country's surface, in the southeast, receives 40 percent of the total amount of rainfall in Mexico (Figure 1). In contrast, 60 percent of the country's population, concentrated in the central regions, receives only 12 percent of the country's rainfall. Almost half of all Mexico's agricultural activities take place in the central regions, and this area supports some of the country's largest urban populations.

More than 70 percent of Mexico's rainfall occurs between May and September (Figure 2), making the country very sensitive to climate variability or climate change processes that are manifest in the spring-summer rainy season. Spring and summer rains are fundamental for the agriculture in the central highlands, an activity that is undertaken basically under rainfed conditions. The sensitivity of rainfed maize to climatic variability and change is of concern because of the economic and cultural importance of this crop in rural areas, particularly for populations that have traditionally depended on their own production for their subsistence.

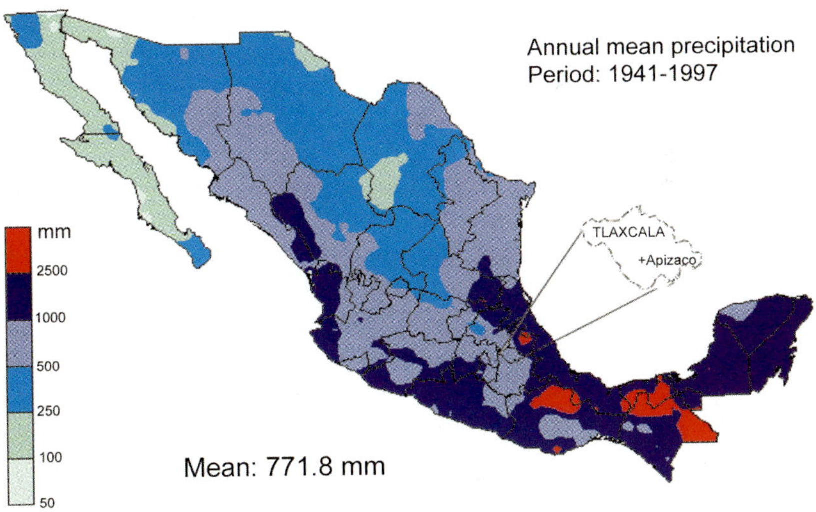

Figure. 1. Annual mean precipitation for Mexico, 1941-1997

Source: National Water Commission, National Meteorological Service, http:/smn.cna.gob.mx/productos/map-lluv/lamina.html.

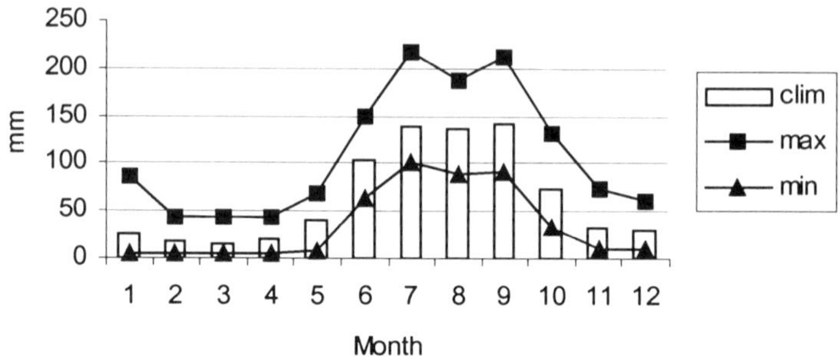

Figure 2. National monthly precipitation averages (in mm; "clim" in the figure), including the maximum and minimum values from 1941 to 1998; "clim" denotes the average of 30 years or more of any climatological variable

2. Previous Research

According to the results of the climate change scenarios developed for the Mexico Country Study (1994-1996) (Gay, 2000), this seasonal sensitivity would be exacerbated under climate change, especially in the central and northern highlands (Magaña et al., 1997). In the country study, we generated climatic base and change scenarios, applying the same methods used by all country studies and generally suggested for these studies (Feenstra et al., 1998). Therefore, the base scenario was constructed using a period of 30 years (1951 to 1980) of data, including temperature (observed, maximum, minimum); precipitation (such as "clim" in Figure 3, which is the average of the 1961-1990 series); and solar radiation at the surface. Monthly and daily data were obtained mainly from the National Meteorological Service (SMN, 2000) and from the Mexican Institute for Technology of Water, Instituto Mexicano de Tecnología del Agua (IMTA, 2000). The climate change scenarios were obtained using the outputs of two general circulation models (GCMs) suggested by the U.S. Country Studies Program (Benioff et al., 1996): the Canadian Climate Center Model (CCCM) and the Geophysical Fluid Dynamics Laboratory (GFDL) model. The monthly anomalies of the basic climatological variables were added to those of the base scenario.

These base and change scenarios conditions were introduced into a model that simulated the crop development under those conditions—the CERES-Maize model (Jones and Kiniry, 1986)—which can also simulate

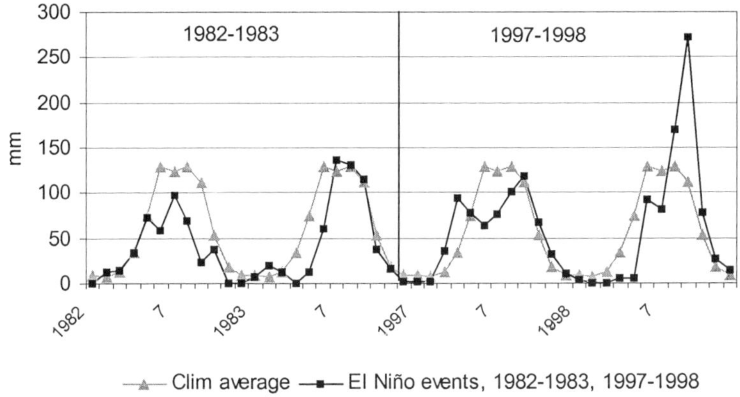

Figure 3. "Clim" indicates monthly average (1961-1990) (base scenario) for Tlaxcala. During 1982-1983 and 1997-1998 El Niño events, lower than normal rain in summer and a delay in the beginning of the rainy season were observed. Also, there was higher than normal precipitation in September and October, particularly in 1998.

the "fertilizing" or physiological effect (PE) responses of maize to an increase or even a doubling of CO_2. Climate change impacts in crop yields were in general negative, except for the sites chosen in the central highlands of the country (Conde et al., 1997), where the model indicated an important increase in crop production, mainly associated with a decrease in the probability of frost. The positive impacts in rainfed agriculture in central Mexico could be quite important for an adaptation strategy in a country where most of the population consumes diverse varieties of maize in their daily diet.

The country study was followed by a case study of climate change in Tlaxcala, which is in the central region of the country (Figure 1). We applied the same methods, confirming the general results that yields would be increasing (Ferrer, 1999). Climatic, crop management and soil data were obtained with the help of the researchers and students from the University of Tlaxcala, particularly for several municipios (counties), such as Apizaco (Figure 1).

For this county (Ferrer, 1999) we chose two different study sites, considering that soil changes are not taken generally into account in the numeric experiments performed with the CERES model, and that soil degradation could eventually occur. For this reason, a site with better soil nutrient conditions was compared with one that was deficient in nutrients. In both cases, the yields increased under the climate change scenarios. Given these results, future increase in yields could be expected, and agricultural planners could look for seeds, fertilizers and other factors that could enhance this possibility. Data related to seed varieties were obtained directly from farmers and from INIFAP, and soil analysis data were obtained from the University of Tlaxcala (Hernández-Rojas, 1991; Rojas-Martínez, 1997). Our optimism over the results of the country study analyses was tempered, however, with observations of recent trends in climatic variability. On average, we found that the overall impact of ENSO events in Mexico tended to reflect the climatic changes anticipated when statistical downscaling techniques were used to generate scenarios, e.g., less rainfall during the summer months and increased precipitation during the winter (Magaña, V., Gay, C., Robinson, R., and Conde C., 1997, project on the use of a climate forecast in planning agricultural activities in the state of Tlaxcala, Mexico; with the support of U.S. Country Studies Management Team and Inter-American Institute; unpublished). The extreme events and local variability associated with ENSO were not, however, consistent with the GCM models. Over the 1990s, in central Mexico, anomalous frost events appeared to be increasingly frequent and precipitation patterns were often irregular, shortening the growing season rather than lengthening it as the models predicted. Moreover, if the changes in climate imply more frequent, intense or larger ENSO events (Watson et al., 2001), the "optimistic" expectations found in these studies could be misleading.

Furthermore, while the CERES-Maize model led us to believe that adaptation through improved crop and farm management would be theoretically possible, during the country study we questioned the feasibility of such adaptation options in the dynamic and often volatile political-economic context of production (Conde et al., 1997). To pursue these questions, we initiated a new phase of research in 1997, this time working with present-day variability and extreme events and actual decision makers in an effort to understand vulnerability and adaptation from the perspective of those most affected by the anticipated climatic changes. Not only did the results of the GCMs give us some confidence in using ENSO events as approximations of the conditions that might become increasingly probable under climate change, but also, by focusing on present-day climatic variability, we were essentially exploring climatic change as the farmers themselves experienced it: as short-term trends, interannual variability and extreme events. We hoped that climate forecasts based on ENSO might prove to be a useful tool in facilitating adaptation not only to extreme events but also to the interannual variability associated with climatic change.

3. Tlaxcala Case Study

We chose Tlaxcala, in central Mexico, as a case study to explore these issues of climatic variability and adaptation (V. Magaña, C. Gay, R. Robinson and C. Conde, 1997 project on the use of a climate forecast in planning agricultural activities in the state of Tlaxcala, Mexico; with the support of U.S. Country Studies Management Team and Inter-American Institute; unpublished) for several reasons. First, the state exemplified the type of agriculture so characteristic of central Mexico: smallholder maize farmers (averaging 3 hectares) planting under rainfed conditions and using low-input technologies. Importantly, the proximity of Tlaxcala to the markets of Puebla, Veracruz and Mexico City meant that despite their technology use and size, these farmers were typically well aware of and affected by changes in agricultural policy and market conditions. Second, the state had been affected by recent ENSO events and thus we anticipated that we would be able to learn which adaptation measures of all the possible ones (found with our models) were adopted by the farmers during these events to overcome the negative ENSO impacts. Finally, and most important, a group of farmers organized in a nongovernmental association named *Fundación Produce AC* were willing to participate in and support this regional project. The project also benefited from the substantial contributions of agroclimatology experts of the University of Tlaxcala. Some of the state's agricultural officials were also interested and were included in the discussion of the results. The participation of these regional actors (stakeholders) turned out to be a fundamental feature in the success and continuity of this experience. The objective of the project was twofold: first, we wanted to understand the climatic variability of the state sufficiently to produce a climate forecast, based on ENSO indicators (V. Magaña, C. Gay, R. Robinson and C. Conde, 1997 project on the use of a climate forecast in planning agricultural activities in the state of Tlaxcala, Mexico; with the support of U.S. Country Studies Management Team and Inter-American Institute; unpublished). Second, we hoped to understand the process of farmers' decision-making and their information needs sufficiently to determine how this information might facilitate adaptation (Eakin, 2000). Over the 1997-1999 period of the study, we had more than 10 workshops with farmers and other stakeholders, and several training mini-workshops. We found that ENSO conditions were a very good starting point to discuss climate perceptions and farmers' climatic needs since our explanation of how local climate variability related to El Niño events and the possibility of regional/local forecasts based on this knowledge were highly appreciated in a community that had not received climatic information on a regular basis from governmental or academic institutions.

We learned in these workshops that to be useful, a climate forecast based on ENSO must include not only the total summer amount of rain but also as much detail as possible about the beginning and distribution of the rainy season. Also, farmers desired information related to hazards like frosts and hails, the two extreme events responsible for their greatest production losses. Moreover, the east-central region of Mexico is affected by a climatic event called the *canícula*, or midsummer drought, characterized by a decline in the amount of precipitation during July and August, depending on the location. If this *canícula* is intense, maize growth can suffer, since the plants are at the end of their silking stage or in the beginning of the filling grain stage. Knowledge of the timing and intensity ("dryness") of the *canícula* was thus understandably important for farmers, particularly for crop choice and seed variety selection.

All of these climatic needs resulted in new research issues for us (Magaña et al., 1999; Morales and Magaña, 1999). From our climatic analysis, we learned that during El Niño conditions, two possible drought situations might appear, suggesting that there might be at least two different strategies for coping with this adverse event. For example, it would be unlikely that the same crop or varieties would be successful in both El Niño years, not only because of differences in the rain's distribution, but also because of the economic losses the farmers experience in the first year of El Niño would be likely to diminish the range of possible choices they have in the second year.

In Figure 3, we show that, during the strong 1982-1983 and 1997-1998 El Niño events, in the first year we might anticipate a normal beginning of the rainy season (compared with the base climatic scenario), but the very strong decrease in the summer rain (a strong *canícula*) might mean important production losses. On the other hand, the second year of a strong El Niño event might imply a delay in the beginning of the rainy season, which would then delay the planting date. This could mean that the early frost in September might appear in the last stages of the plant's growth. Also, higher than normal precipitation in September-October in the second year (particularly in 1998) could exacerbate erosion processes and lead to more crop losses.

The variation in the timing of the initial rains and the nature of the *canícula* during different phases of the ENSO event underline the importance of the farmer's questions: "Is the *canícula* coming with water?" or "Is it going to be a dry *canícula*?" These were questions that we could not answer until we studied the 1997-1998 event. We now recognize that a useful regional forecast must include the "dryness" or "wetness" of the midsummer drought.

The severe impact of anomalous frost events on farmers' yields in the late 1990s and the farmers' interest in frost information as part of their agricultural planning led us to explore in more detail the implications of ENSO for frost. The results from runs of the CERES-Maize model illustrated that if future climate conditions were expected to be similar to an El Niño event, then the possible positive impacts expected from the GCMs (Ferrer, 1999) would be severely diminished. This would be because of the increase in early frosts during El Niño events, which could be associated to a general decrease in atmospheric humidity conditions (radiative frosts). The longer growing season and less frequent frosts associated with the GCM outputs were the opposite of what farmers had experienced and what we modeled with the CERES-Maize model for ENSO conditions (Conde et al., 1999) (Figures 4a and 4b).

Once we understood the importance of climate variability and extreme events such as frost to the farmers we consulted, our next step was to understand the feasible options for adaptation to such events. Agricultural adaptation strategies suggested in the climate change literature, and reflected in analyses involving crop models, typically involve adjustments in crop management (input applications, planting/tillage methods and timing) and crop and seed choice, as well as a variety of other adaptations involving accessing new markets, use of insurance or production contracts and other mechanisms (Easterling, 1996).

In Tlaxcala, the experience of the farmers of *Fundación Produce A.C.* with the 1998 forecast led us to believe that one of the more promising options for adaptation was crop choice. The farmers in the association

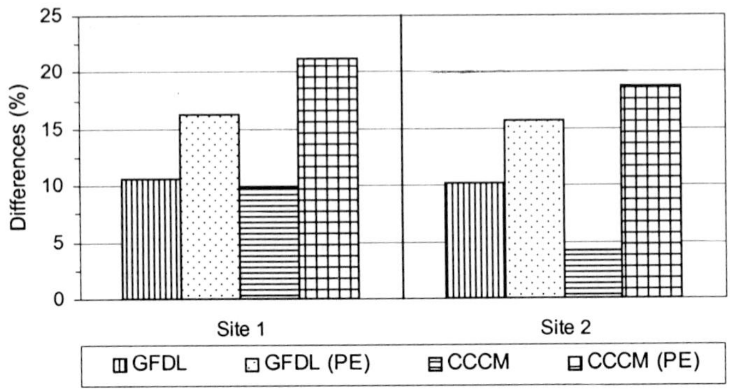

(a)

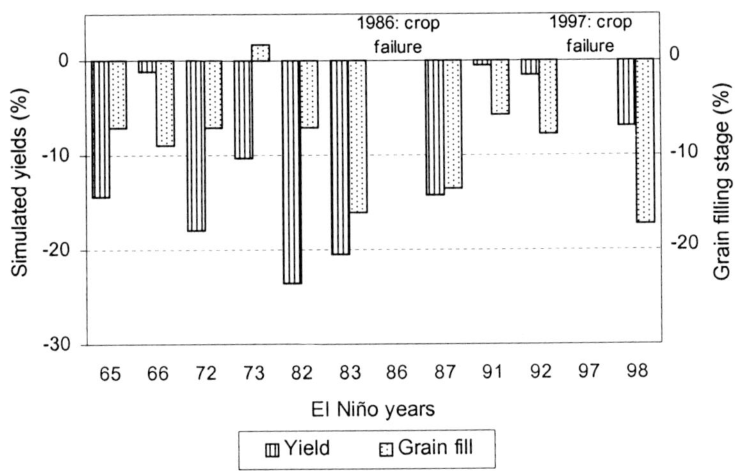

(b)

Figures 4a and 4b. (a) Results obtained with the CERES model for two study sites show increases in the yields in Apizaco, Tlaxcala, under climate change scenarios (GDFL and CCCM) with respect to observed (OBS.) and climatic baseline scenario (BASE). PE refers to the physiological effect of a doubling of CO_2 on maize. (b) Under El Niño climate conditions, the CERES model, evaluated only in Site 1, reports decreases in the yields specifically as a result of a decrease in the filling grain stage of maize.

decided in May to make an early (and thus less expensive) purchase of oat seeds, basing their decision on the regional forecasts we delivered and on the crop suggestions and financial support of the government agencies with whom we were working. The harvest of these farmers was used mainly as forage for their cattle, and as seed for the subsequent years' planting. This application of the 1998 forecast was one of the most successful achievements of our climatic research, suggesting that the forecasts could have an important function in cropping decisions in the state. Nevertheless, we realized that this was a partial success obtained with a small group of farmers in a very strong ENSO event. Our next challenge was to understand the process of agricultural decision-making in greater depth, particularly if we were to hope that forecasts would have utility for a broader segment of the agricultural population and under circumstances of greater climatic risk.

4. Adaptation to Climatic Variability in Tlaxcala

The success of the farmers' oat crop under the 1998 climatic conditions was not completely unexpected. In workshops and interviews, farmers suggested that with the expectation of a late onset to the rainy season, they would consider annual adaptation measures ranging from using different varieties of maize seeds with shorter development cycles to switching from planting maize to planting shorter-cycle crops like barley, wheat or oats (Eakin, 1998). After analyzing the suitability of different crops for the state under ENSO conditions, the government of Tlaxcala was also promoting these crops as alternatives to maize (Legorreta-Padilla, 1998). Oats were considered particularly advantageous because of their 80-day growing cycle compared to the 130 to 160 days for most of the state's maize varieties.

To facilitate this "crop conversion", the agricultural ministry of Tlaxcala was offering small quantities of hybrid barley and oat seed as well as commercial maize seed at a discount through a program called *Kilo por Kilo,* with the hope that farmers would gradually replace local maize varieties with more crops considered more appropriate both in climatic and commercial terms (Legoretta-Padilla, 1998). Oat seeds were also distributed free to farmers who reported the heaviest losses to frost in 1999. The production volume of barley and oats in Tlaxcala has increased in recent years, perhaps in response to persistent losses in maize and the government's promotion of alternative crops (Figure 5).

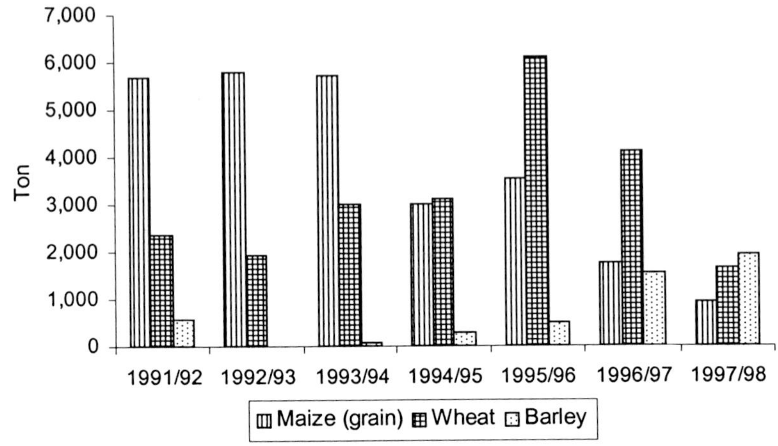

Figure 5. Maize, wheat and barley production for Apizaco, Tlaxcala (1991-1998). The trends shown can be associated with farmers' adaptation responses to the climatic extreme events during those years and also to the agricultural policy in the region to support changes in crop production.

To explore in greater depth the opportunities for and obstacles to adaptation in Tlaxcala, ethnographic research involving surveys and in-depth interviews was conducted with households in two rainfed agricultural communities in Tlaxcala in 2000 (Eakin, 2000a). The communities were representative of the two predominant types of smallholder producers in the state: those that are primarily oriented toward subsistence production and those that are "semi-commercial", selling a variety of rainfed grains in small quantities in local markets. The interviews revealed that most farmers were quite aware of the adaptive advantages of oats and barley in comparison to maize in years when the onset of the rainy season was delayed or the risk of frost was perceived to be high. Despite this recognition, the household data also suggested that persistent uncertainties and risk related to participating in both consumer and producer markets might impede some households from switching crops as an adaptation strategy (Eakin, 2001b).

Until the mid-1980s, farmers had enjoyed price supports for all basic grains as well as subsidized fertilizer and seed from state-owned agencies PRONASE and FERTIMEX. For Mexico's smallholders, this support was typically channeled through the public agricultural bank, BANRURAL.

Although the bank's influence over farmer's production choices and technology was criticized as "patronizing" and heavy-handed, credit for smallholders was relatively easily acquired, and each loan was accompanied by mandatory insurance (Myhre, 1998). In the aftermath of the 1982 economic crisis, these institutions were rapidly dismantled. By 1989, price supports were withdrawn for all crops but maize and beans, and the public input suppliers were closed or privatized credit (de Janvry et al., 1997). BANRURAL was restructured to focus on "commercially viable" farmers, such that between 1989 and 1991 the land accredited by BANRURAL fell from 7.2 million hectares to 1.2 million hectares, essentially excluding the vast majority of Mexico's smallholders from formal credit (de Janvry et al., 1997; Myhre, 1998). By 1996, less than 2 percent of Mexico's maize area was insured (Myhre, 1998). While few would argue reform was needed, in the late 1990s, farmers were left in what de Janvry et al. had called "an institutional vacuum" (de Janvry et al., 1995).

Partly as a result of these changes, the market for barley and oats is now quite narrow. Both barley and oats are commercial crops and are not processed by the households for human consumption. Farmers have neither a guaranteed price nor a guaranteed buyer for their harvests. Farmers can attempt to sell their harvests to the malt industry for a potentially higher price than they would receive in local markets, but unless they invest in the seeds and chemical inputs required to meet the industry's standards, they are unlikely to be able to access the market. Even with the *Kilo por Kilo* subsidy, this investment is risky, for the very reason that frost events are frequent and insurance is inaccessible. It is far safer to minimize investment costs by planting from local seed stocks, although this strategy restricts a farmer to selling the harvest as grain, forage or hay to local merchants and seed stores. In these markets, the prices are rarely better than those for maize (Figure 6), and at the relatively low yields obtained by farmers in Tlaxcala (averaging just over 2 tons/hectare for barley), few farmers can guarantee that they will earn enough from their sales to purchase the maize they need for their own consumption.

The household surveys and interviews revealed that those households who planted oats or barley were planting these crops in addition to maize rather than "switching" out of maize entirely. Those households who reported planting oats in 1998 and 1999 also reported fewer crop losses in those years, confirming the farmers' perceptions of the climatic advantages of the crop. However, despite the widespread recognition of these

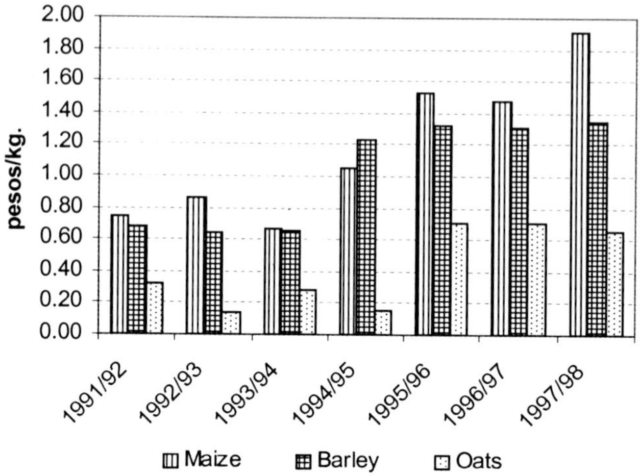

Figure 6. Prices of basic grains in Tlaxcala, from 1991 to 1998. Among other features, price of maize is also an important element for farmers' choices.

Source: Agrit Database, SAGAR-Tlaxcala.

advantages, fewer than half of the surveyed households in the two communities planted oats. Those that did were households with more than 3 hectares; they planted oats in addition to several hectares in maize, substantiating farmers' comments that only when their household food needs were met would they consider diversifying their crop mix.

Given the obstacles for commercialization of oats and barley, those households who had sufficient land to diversify their crop base planted oats and barley only if they had sufficient livestock to consume the harvests as forage. Milk cows and goats were the preferred investments in the two communities. Although the markets for small-scale livestock production are not well developed in Tlaxcala, like maize, the animals provided the households with both subsistence products (milk and meat) and cash to facilitate the recuperation of crop losses.

These case studies suggest that land area and subsistence requirements may be an obstacle to the adoption of an oat/barley/livestock strategy, either with or without the assistance of a forecast. Despite the sensitivity of maize to precipitation anomalies and frost, for smaller-scale households maize represents food security and insurance against economic uncertainty. Rural households can use maize for their own subsistence as well as for their livestock, and, in comparison to other grains, maize grains can more easily be cooked, stored and sold. Farmers also still control knowledge of organic pest management and fertilizers, and other management features (Warman, 1995). Neither oats nor barley are consumed directly by households, thus unless a household has large livestock, these crops are far less useful (Eakin, 2001b). Thus, while the combination of economic changes and a decade of adverse climatic conditions has encouraged more farmers to adopt barley and oats in Tlaxcala (Figure 5), our case studies in the state suggest that this adoption may be only *in addition* to maize and only with those households where sufficient land is available to diversify into alternative grains (Eakin, 2001b).

5. Conclusions

Our research in Tlaxcala illustrated the importance of involving key regional actors in our exploration of the nature of climatic variability, its impacts and the potential for adaptation. The success of our analysis of climatic variability and trends was in part due to the active participation of farmers, government officials and agroclimatic researchers in the state in the discussion of climate impacts and their importance to particular aspects of agricultural decision-making. Through these discussions we identified priorities for our research, and came to recognize the importance of locally held perceptions and knowledge of climate in the design of a forecast. We not only were able to conduct research that was more appropriate to the region, but we also formed the basis of a common language in which climatic uncertainty and risk could be discussed with stakeholders. Our resulting analysis of the relationship of ENSO to local climatic features illustrated the potential utility of seasonal climate forecasts for decision-making.

Translating such analysis of variability and impacts into a practical tool for adaptation has proven more challenging than we anticipated. While the farmers who participated in the project were fully aware of the potential utility of forecast in seed and crop choice decisions, ethnographic research in several communities in Tlaxcala revealed that such presumably simple adaptation strategies are far more complex than they would seem. Households are constantly in a process of adaptation, but not only to climate variability and change. Rural producers are also simultaneously adjusting to changing institutional circumstances and opportunities that affect the viability of their subsistence and household economy. These nonclimatic factors play significant roles in their strategies such that simulated "optimal" adaptations such as planting oats instead of maize during El Niño years may in fact do little to achieve household goals of food security.

There are obvious advantages to planting alternative crops during years where anomalous rainfall is predicted. The increase in the production of oats and barley in the latter half of the 1990s illustrates the response of some farmers to the promotion of these crops. Yet our interviews with farmers revealed that despite the strong promotion of barley and oats, maize—particularly local varieties—remain preferred crops given their role in the household economy and the stability that subsistence maize offers at a time when consumer prices and input costs are escalating. This is particularly true for those farmers for whom planting alternative crops would mean forgoing area for maize. Thus farmers continue to plant a wide variety of maize, even in very small amounts, for use in several annual festivities; this allows them the flexibility the different varieties offer. From the point of view of adaptation, maize production is still the activity that gives the small farmers a broader set of choices (Warman, 1995) and more flexibility in face of both economic and climatic risk. Our findings indicate that if climate forecasts are to be associated with crop choice in Tlaxcala, they will be most effective in a political-economic environment that is supportive of the commercialization of these alternative crops.

It also appears that for smaller-scale farmers, the development of commercial opportunities for shorter-cycle maize varieties may also facilitate adaptation. Farmers are already accustomed to planting rapidly maturing local landraces along with their higher-yielding white maize varieties. There has been little attention in Tlaxcala to developing commercial opportunities for these alternative maize varieties for those households who are reluctant to abandon maize as their primary crop. While further research on the potential of climate information more generally and climate forecasts specifically to facilitate adaptive responses in agriculture is important, our research has illustrated that considerable attention must be paid to the broader institutional context in which farmers operate. As has been argued elsewhere, understanding the process of adaptation requires understanding risks and uncertainties from the perspective of the decision maker. Not only, as has been argued elsewhere, does this mean explicit attention to climatic variability and extreme events (Burton, 1996; Smit et al., 1996), but it also means a concerted attempt to account for the dynamic social environment that in many circumstances may represent to the farmers a far greater source of uncertainty than changes in climate (Chiotti et al., 1997; Smithers and Smit, 1997).

Acknowledgments

We want to thank R.M. Ferrer for her help and comments on the CERES model results and Joel Smith and Ana Iglesias for their valuable suggestions to improve this work. H. Eakin would also like to acknowledge the financial support of a National Science Foundation Graduate Fellowship for the social science research component of this project, and the support and patience of the farm households who participated in this study.

References

Benioff, R., Guill S. and Lee J., eds. (1996), *Vulnerability and Adaptation Assessments: An International Handbook.* U.S. Country Studies Management Team. Kluwer Academic Publishers, The Netherlands.

Burton, I. (1996), The growth of adaptation capacity practice and policy. In *Adapting to Climate Change: Assessments and Issues,* Smith, J., Bhatti, N., Menzhulin G. et al., eds. Springer, New York, 55-67.

Chiotti, Q. et al. (1997), Agricultural response to climate change: a preliminary investigation of farm level adaptation in southern Alberta. In *Agricultural Restructuring and Sustainability,* Ilbery, B., Chiotti Q. and Rickard. T., eds. CAB International, Oxon and New York, 201-218.

Conde, C., Ferrer, R. and Liverman D. (2000), Mexico. Una vision hacia el siglo XXI. El Cambio Climático en México. Resultados de los Estudios de Vulnerabilidad del País, Coordinados por el INE con el Apoyo del U.S. Country Studies Program, 119-141 SEMARNAP, UNAM, Mexico.

Conde, C., Magaña, V. and Ferrer, R. (1999), Paper presented at the 11th Conference on Applied Climatology, 10-15 January, Dallas, TX, USA.

Conde, C., Liverman, D., Flores, M., Ferrer, R., Araujo, R., Betancourt, E., Villarreal, G. and Gay. C. (1997), Vulnerability of rainfed maize crops in Mexico to climate change. *Climate Research* **9**, 17-23.

De Janvry, A. et al. (1995), Reformas del sector agrícola y el campesinado en México. Fondo Internacional de Desarrollo Agrícola e Instituto Interamericano Cooperación para la Agricultura, San José de Costa Rica.

De Janvry, A. et al. (1997), Mexico's Second Agrarian Reform: Household and Community Responses. San Diego Center for U.S.-Mexico Studies, University of San Diego, CA, USA.

Eakin, H. (1998), Adapting to Climate Variability in Tlaxcala, Mexico: Constraints and Opportunities for Small Maize Producers. Geography and Regional Development. M.A. Thesis. University of Arizona, Tucson, AZ, USA.

Eakin, H. (2000), Smallholder maize production and climatic risk: a case study from Mexico. *Climatic Change* **45**, 19-36.

Eakin, H. (2001a), Paper presented at the Annual Meetings of the Association of American Geographers, February 28- March 3, New York.

Eakin, H. (2001b), Paper presented at the Open Meetings of the Human Dimensions Research Comm., October 6-8, Rio de Janeiro, Brazil.

Easterling, W. (1996), Adapting North American agriculture to climate change in review. *Agricultural and Forest Meteorology* **80**, 1-53.

Feenstra, J., Burton, I., Smith, J. B. and Tol, R. S. J., eds. (1998), *Handbook on Methods of Climate Change Impacts Assessment and Adaptation Strategies.* UN Environment Programme, Nairobi, Kenya, and Institute for Environmental Studies, Amsterdam, The Netherlands.

Ferrer Perdomo, R. M. (1999), Impactos del cambio climático en la agricultura tradicional del maíz en Apizaco, Tlaxcala. tesis. Universidad Nacional Autónoma de México.

Gay García, C., compilador (2000), México: Una Visión hacia el Siglo XXI, El Cambio Climático en México. Resultados de los Estudios de Vulnerabilidad del País, Coordinados por el INE con el Apoyo del U.S. Country Studies Program. SEMARNAP, UNAM, México.

Hernández-Rojas, F. C. (1991), Sistemas y Técnicas Agrícolas Tradicionales en Tlaxcala. Gobierno del Estado de Tlaxcala, Mexico.

IMTA (2000), ERIC II (Extractor Rápido de Información Climática). CD edited by Instituto Mexicano de Tecnologia del Agua, México.

IPCC , WGI (2001), Summary for Policy Makers. A Report of Working Group I of the Intergovernmental Panel of Climate Change.

Jones, C. A. and Kiniry, J. R. (1986), *CERES—Maize: A Simulation Model of Maize Growth and Development.* Texas A&M Press, College Station, TX, USA.

Legorreta-Padilla, F. (1998), Conversión productiva del Temporal para Tlaxcala. Opciones para reducir el impacto de Eventos climáticos en el Agro. SAGAR, AGROTLAX e INIFAP. Fundación Produce, Tlaxcala, México.

Magaña, V., Amador, J. and Medina, S. (1999), The midsummer drought over Mexico and Central America. *American Meteorological Society* **12**, 1577-1588.

Magaña, V., Conde, C., Sánchez, O. and Gay C. (1997), Assessment of current and future regional climate scenarios for Mexico. *Climate Research* **9**, 107-114.

Morales, T., and Magaña, V. (1999), Unexpected frosts in central Mexico during summer. *Proceedings of the 10th Symposium on Global Change Studies, 10-15 January, 1999, Dallas, TX, Preprint Volume.* American Meteorological Society, Boston, MA, USA, 262-263.

Myhre, D. (1998), The Transformation of Rural Mexico. Reforming the Ejido Sector. W. Cornelius and D. Myhre. San Diego Center for U:S.-Mexico Studies. University of San Diego, San Diego, CA, USA.

Reyes Castañeda, P. R. (1981), Historia de la agricultura: Información y Síntesis. México. AGT Editor, S.A., México.

Rojas-Martínez, I. (1997), Guía para la Producción de Maíz de Temporal en Tlaxcala. Folleto Técnico No. 4. SAGAR, AGROTLAX, INIFAP. Fundación Produce, Tlaxcala. México.

SEMARNAT (2001), Crusade for Forests and Water. Minister of Environment and Natural Resources. Available at http://www.semarnat.gob.mx/bosque-agua/ vision.html. (Spanish).

Smit, B. et al. (1996), Agricultural adaptation to climatic variation. *Climatic Change* **33**, 7-29.

Smithers, J. and Smit, B. (1997), *Agricultural Restructuring and Sustainability. A Geographical Perspective.* Ibery, B., Chiotti, Q. and Rickard, T., eds. CAB International, Oxon and New York.

SMN (2000), Dat 322 V.1.0. Gerencia de! Servicio Meteorológico Nacional. Developed by Instituto Mexicano de Tecnologia del Agua, México.

Warman, A. (1995), *La Historia de un Bastardo: Maíz y Capitalismo*. FCE, UNAM, Mexico.

Watson, R. T. et al. (2001), Summary for policy makers. In *Climate Change 2001: Synthesis Report, Contribution of Working Groups I, II, and III to the Third Assessment Report of the Intergovernmental Panel on Climate Change,* Watson, R. T. and the Core Writing Team. Cambridge University Press, Cambridge, UK, 1-34.

12. Engaging Africa in Adaptation to Climate Change

Chris H. D. Magadza
University of Zimbabwe
Harare, Zimbabwe

The Third Assessment Report of the Intergovernmental Panel on Climate Change identified Africa as the continent most vulnerable to anthropogenic climate change. The vulnerability of Africa stems from its low adaptive capacity. Against the backdrop of projected change climate impacts, this paper notes the lack of definitive commitment to climate change adaptation in Africa, despite measurable national costs attached to climate related events. In Africa, climate change impacts are considered against a context of degrading environment and underperforming agricultural systems. The paper poses the problem of the time phase difference between the manifestation of climate change issues and need for adaptation strategies on the one hand, and the development of concepts of good governance in Africa on the other. While strengthening the capacity of African governments to create enabling legal and institutional environments for climate change adaptation initiatives is important, enabling the private sector through relevant technology transfer, entrepreneurship enhancement and operation of market forces is likely to be the most productive strategy in climate change adaptation. The Zimbabwe situation is used throughout the paper as an illustrative study.

1. Introduction

Africa has lagged behind the rest of the world in development. The lowest gross domestic product (GDP), highest infant mortality, highest nutritional deficit, declining per capita access to adequate and safe water supplies and declining economic performance are to be found among African countries. The 10 poorest countries in the world are African. Because of a combination of population growth and dependence on primary natural resources for energy and food production, the African environment is experiencing some of the highest rates of degradation in the world. Poor sanitation and limited health care are reflected in the high levels of debilitating illnesses (UNEP, 2000).

Now climate change threatens to further compound this situation with its projected impacts on the African environment and its people. Because of its low development status, Africa's capacity to adapt to the impacts of climate change is low, but of even more concern is the lack of attention from African leaders to the future harm to their nations from climate change impacts.

However, Africa is not a homogenous continent. Its nations vary both in degrees of development and in political development stages, and so generic statements are not equally applicable to all states. A notable exception, for example, to the generalities listed above is Botswana, which has experienced political stability and consistent economic growth since independence in 1966, in spite, ironically, of being one the arid countries of sub-Saharan Africa. North Africa showed steady economic growth in the decade between 1975 and 1985, but that growth has since slowed. Southern and West Africa show the most pronounced and consistent economic decline.

2. Impacts Overview

The Intergovernmental Panel on Climate Change (IPCC) Third Assessment Report (TAR) (McCarthy et al., 2001) identified Africa as the continent most vulnerable to the effects of climate change. Even without the added impacts of climate change, most of Africa outside the equatorial belt is projected to experience water supply stress by the end of the first quarter of this century, through a growing imbalance between supply and demand. According to the IPCC TAR, precipitation in Africa will continue to be variable, but current semiarid and arid lands generally will be drier, with runoff projected to diminish by between 5 percent and 40 percent (Table 1). The synoptic view of freshwater availability (UNEP, 2000) indicates that North Africa, The Horn of Africa, Malawi, the Great Lakes states of Rwanda and Burundi and South Africa will experience severe water scarcity, while much of central West Africa, East Africa and Zimbabwe will experience water stress, Table 2 shows likely changes in water availability by 2025 (Gleick, 1993). Water stress situations have already begun to manifest as potential sources of conflict in riparian states in shared water basins such as the Nile and the

Table 1. Estimates of ranges of percentage changes in precipitation, potential evaporation and runoff in African river basins

Basin	Change in precipitation (%)	Change in potential evaporation (%)	Change in runoff (%)
Nile	10	10	0
Niger	10	10	10
Volta	0	4 to -5	0 to -15
Schebeli	-5 to 18	10 to 15	-10 to 40
Zaire	10	10 to 18	10 to 15
Ogooue	-2 to 20	10	-20 to 25
Rufiji	-10 to 10	20	-10 to 10
Zambezi	-10 to -20	10 to 25	-26 to -40
Ruvuma	-10 to 5	25	-30 to-40
Limpopo	-5 to -15	5 to 20	-25 to -35
Orange	-5 to 5	4 to 10	-10 to 10

Source: Based on McCarthy et al., 2001, Chapter 10.

Kovango. Nicholson (1998) draws attention to lake level variations in some Rift Valley lakes, which could have far-reaching impacts on the economies associated with them.

For over four decades, Africa's food production has been declining (Chapter 2 in UNEP, 2000). The major cause of this decline has been land degradation coupled with deteriorating climatic conditions. This decline has been accompanied by an increasing frequency of famines. The IPCC TAR projects diminished crop yields because of insufficient precipitation, especially in the Sudano Sahel and Southern Africa. The possible impact of climate change on maize production in Zimbabwe was evaluated by simulating crop production under climate change scenarios generated by global climate models (GCMs) (Muchena and Iglesias, 1995). Temperature increases of 2°C or 4°C reduced maize yields at all sites, even when the beneficial fertilization effects of CO_2 were included.

There are increasing reports of coastal instability (UNESCO, 1997), placing much infrastructure and service facilities at risk. Coastal subsidence due to drops in water tables in delta areas, coastal erosion due to environmental degradation in coastal zones and flooding due to sea level rise are the major threats to Africa's coastal environment (IPCC, 1998). Victoria

Table 2. Water availability for some Africa countries, 1990 and 2025 (m^3 per year)

Country	Per capita availability 1990	Per capita availability 2025
Algeria	750	380
Burundi	660	280
Cape Verde	500	220
Comoros	2040	790
Djiboute	750	270
Egypt	1070	620
Ethiopia	2360	980
Kenya	590	190
Lesotho	2220	930
Libya	160	60
Morocco	1200	680
Nigeria	2660	1000
Rwanda	880	350
Somalia	1510	610
S. Africa	1420	790
Tanzania	2780	900
Tunisia	530	330

After Gleick, 1993.

Island, an area of highly valued real estate in Lagos, recently suffered severe flooding. Hundreds of homes were destroyed by Nile flooding in Khartoum during the floods of 2001. The IPCC TAR warns of severe consequences to some coastal areas because of sea level rise. According to model results, coastal cities such as Banjul in The Gambia could lose as much as 45 percent of their present area to sea level rise and subsequent coastal erosion (Jallow et al., 1999). Box 1 describes the population and economic activity in coastal West Africa (UNEP, 2000), which suggests that a substantial proportion of the population and capital investments in western African states is potentially threatened by changes in sea level. Other possible impacts to human settlements in such coastal areas in Africa are discussed in Magadza (2000).

Box 1. Coastal population and economic activities in West Africa

A large percentage of West Africa's urban population lives in coastal cities. In Nigeria, for example, about 20 million people (22.6 percent of the population) live along the coastal zone; about 4.5 million Senegalese (66.6 percent of the population) live in the Dakar coastal area. About 90 percent of industry in Senegal is located within the Dakar coastal zone. In Ghana, Benin, Togo, Sierra Leone and Nigeria, most of the economic activities that form the backbone of the national economies are located within the coastal zone. Coastal areas also form the food basket of the region. Offshore and inshore areas as well as estuaries and lagoons support artisan and industrial fisheries, which account for more than 75 percent of fishery landings in the region (UNEP, 2000).

Other areas of human well-being threatened by climate change are human and animal health, ecosystem integrity and biodiversity. Indeed, the TAR warns of possible extinction of species and irreversible ecosystem changes that could have wide impacts on sustainable development in, for example, the ecotourism industry.

Extreme events of droughts and floods in eastern and southern Africa have been linked to the sea surface temperature anomalies (SST) and the El Niño/La Niña phenomenon of the Pacific Ocean, while those in West Africa have recently been associated with the Atlantic dipole (see Marshall et al., 2001). Figure 1 shows the decadal changes in frequency of the SST anomalies of the western Pacific. It can be seen that the frequency of above average anomalies, generally associated with droughts in southern Africa, has changed during the last two decades. In particular, anomalies exceeding one normal deviate, absent from the series during the decades before 1978, have appeared with increasing frequency in the last two decades. These trends point to a probable increase in the frequency of extreme weather events in the short to medium time horizons, a conclusion also arrived at by the IPCC.

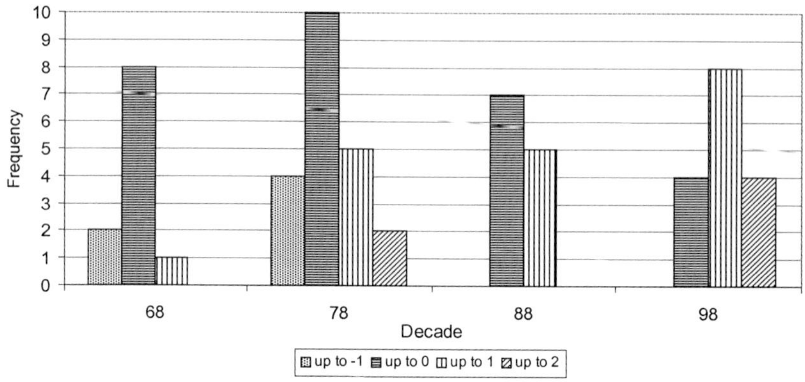

Figure 1. Decadal shifts in El Niño intensity frequencies. Note the disappearance of the La Niña episodes after the 1988 decade.

The IPCC Second Assessment Report (IPCC, 1996) collated information that indicates possible shifts in both the frequency and the occurrence of pathogen induced human diseases. For example, an observed increase in the frequency of malaria incidences in the highlands Rwanda has been linked to changes in rainfall and temperature. In vector borne diseases, climate affects distribution patterns of vector and pathogens through its effects on their life cycles. In cooler climates, pathogens may exist without transmissivity since the temperature required for pathogen development is generally higher than required for the vector (O'Malley, 1992).

Many parts of Africa are protected from vector borne diseases by either elevation or latitude. The MARA/ARMA (1998) project has modeled malaria distribution in Africa. The distribution map shows islands of malaria free areas embedded in malaria risk areas, such as the high veldt of Zimbabwe, as well as areas that are free because they are too removed from the tropics. The project, however, made no attempt to map these distributions to potential future climate. Using the MAGICC/SCENGEN model (Wigley et al., 2000) it is possible to show that the suitable areas for stable malaria transmission will migrate southward at about 1.65° latitude per 25 years in low-lying areas to 0.35° latitude on higher ground. It is also possible to show that areas currently free from the disease because of elevation will be malaria risk areas by 2025.

However, the MARA/ARMA document notes that the presence of malaria will be due to a combination of factors that can be summarized by the human development index (HDI), a composite index of development consisting of measures of life expectancy, income and educational attainment (UNEP, 2000). The predisposition of Africa to climate change induced health risks stem from poor development of state services in health care, environmental management and land use. These observations on malaria can be applied to similar vector borne diseases such as sleeping sickness, dengue fever and schistosomiasis, with specific considerations for their biology.

Africa derives considerable revenue from ecosystem services such as biodiversity, which supports the tourist industry. Again, the IPCC noted the potential adverse impacts on biodiversity in Africa due to a combination of climate change and land use. The latter arises from foreclosure of ecosystem adaptation through biome shifts because of habitat fragmentation. In the land reform program of Zimbabwe, subsistence farmers have taken up residence in wildlife management areas such as the Gonorezhou National Park, thus prejudicing the tourism potential in that area as well as scuttling an international program between South Africa, Mozambique and Zimbabwe to create an interconnected conservancy area in the subregion.

In summary, the African continent faces climate change related threats in

- water supply
- food production
- coastal settlements
- water borne and vector borne disease
- ecosystem changes and biodiversity loss.

3. Current Costs of Climate Related Events

Although no systematic costing of extreme events is available for most of Africa, the floods and droughts in various parts of Africa in the 1999 season provide a window on the magnitude of the costs of such events (OCHA, 2000). The cost of Cyclone Eline in relief in Zimbabwe amounted to US$2,556,179 in donations. The Nigerian government has estimated the cost of neglecting environmental issues. The cost due to coastal erosion is approximately US$150 million annually. Soil degradation alone has an annual cost of nearly 9 percent of GDP, while coastal erosion accounts for

0.5 percent of GDP in that country, but may also climb into the single digit figures within a few decades (World Bank, 1990).

Table 3 shows the cost of relief operations related to drought and flooding in Africa in 1999 (from United Nations, 1999). The rehabilitation costs for the Mozambique floods was over $600 million. The total for Africa in that year accounted for nearly 50 percent of the total UN disaster relief operations. Africa is currently the main recipient of world relief assistance for climate related disasters.

The costs listed in Table 3 relate only to relief operations. Climate related events are beginning to affect development and industrial productivity. Box 2 illustrates the possible magnitude of the economic impacts of climate variability in one drought cycle in the early 1990s in the Zambezi hydrological region. Similar curtailment of energy production due to drought has been reported in the media in West Africa and in Kenya, though no cost estimates are available. It is worth noting though that the Akasombo facility on the Volta River in Ghana is a regional powerhouse, and thus energy production curtailment at this power plant due to climate change has regional implications.

Table 3. Relief agency costs associated with floods and droughts in several countries in Africa in 1999

Country	Cause	Amount (US$)
Chad	Flood	50,000
Democratic republic of the Congo	Flood	1,270,850
Djibouti	Drought	1,555,582
Ethiopia	Drought	5,263,501
Gambia	Flood	145,000
Ghana	Flood	1,317,525
Mozambique	Flood	1,523,143
Togo	Flood	146,853
Total		11,272,454

Data derived from United Nations, 1999.

Box 2. Costs of the 1991/1992 drought due to a reduction in power generation on Lake Kariba

"Economic impacts from curtailment of the hydropower generation from Lake Kariba, resulting from the 1991/1992 drought, were estimated to be some US$102 million loss in GDP, US$36 million loss in export earnings and loss of 3000 jobs (Benson and Clay, 1998). The direct impacts on agriculture and the knock on impacts were also quite severe. These limited estimates provide a window on possible economic impacts of climate change mediated water resources changes in the medium term, i.e., into the middle of this century, a time span within the planning window of economic development strategies" (Desanker and Magadza, 2001, p. 499).

As Table 3 indicates, climate related disasters in Africa attract considerable relief operations. These relief costs are not reflected in national budgets, and thus national accounts do not reflect the mounting costs of disasters related to climate change and climate variability. One of the major findings of the TAR is the possible increase in extreme events such as severe drought, flooding or sea surges in coastal zones. Increases in the frequency of these events could cause considerable financial drain, especially when the ensuing costs constitute a significant portion of a nation's GDP. Thus, failure to adapt to recurring extreme events can negate the entire development process.

4. Response of African States to Climate Change Threats

In spite of these impending threats to sustainable development by anthropogenic climate change impacts, acting over and above general degradation of the natural resource base (desertification, loss of forest, siltation of rivers, coastal erosion), there is no coherent response by African governments to issues of climate change adaptation needs.

In the cycle of the preparation of TAR, a draft report of likely impacts of climate change to the African continent and possible adaptation strategies was submitted for government review and comment. Governments from outside Africa provided substantive responses, but there were no responses from the African governments. The participation of most African countries in climate change activity has largely been confined to meeting their obligations for the periodic production of the country communications, mainly on emissions of greenhouse gases, in accordance with Article 4(a) of the UN Framework Convention on Climate Change (UNFCCC). Few, if any, African countries have embarked on proactive programs aimed at mitigating the impacts of climate change or adapting to them.

An illustration of this failure to incorporate climate change threats in development planning is the proposed Batoka hydroelectric plant on the Zambezi. In the design of the planned Batoka Gorge hydroelectric dam, as in the design of other reservoirs in Africa, the planners used historical flow data as a design parameter of the future reservoir. However, climate change projections indicate significant runoff reduction on the Zambezi (Magadza, 1996). At various stages it was also proposed to increase the generation capacity of Lake Kariba on the Zambezi by installing more generating units, in spite of the demonstration in the 1991/1992 drought that the live storage of the dam can fall short of demand during such droughts, a situation predicted by Urbiztondo (1992). Experts' comments on the omission of climate considerations in this important investment were ignored.

Besides the proposed hydroelectric scheme, water supply constraints in major urban areas in the region (Bulawayo, Johannesburg, Gaberone, Windhoek) have prompted consideration of the Zambezi as a regional water supply facility (e.g., Mapande, 1995). Again, no climate change considerations are factored into such plans.

Similarly, in the area of food security, the mismatch between climate change concerns and government perspectives can be illustrated by the Zimbabwe situation. The IPCC TAR notes that climate change impacts on food security can be exacerbated by no sustainable land use. In Zimbabwe's so-called communal lands, as in many African countries, land users have neither authority nor accountability for the land they use. This system has barred investment and entrepreneurship in sustainable land use for well nigh a century, leading to sustained land degradation, which compromises the productivity of the land. Figure 2 shows the disparity in production levels between communal land use system and commercial farming in Zimbabwe. While the figure illustrates a large disparity in productivity between the two land use systems, it also shows how, in drought periods, communal land productivity drops to near zero, while the capital intensive commercial farming system could still register harvests of over two tonnes per hectare.

The quote in Box 3 is an illustration of the common divergence between attempts to rationalize the environment management process and political

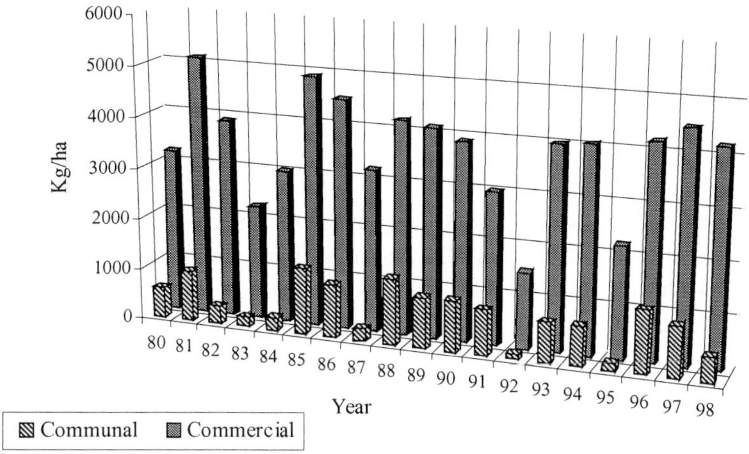

Figure 2. Food production responses to climate variability in commercial and communal land agrarian systems in Zimbabwe. Data reconstructed from FAO (2000).

perspectives. It is taken from the Zimbabwe assessment of climate impacts on agriculture, conducted by the state sponsored Scientific Industrial and Research Development Centre (SIRDC) (Matarira et al., 1995). In contrast to the recommendation in the report, the government is engaged in an economically and environmentally disastrous land redistribution program as a response to the emergence of an opposition polity. Figure 2 suggests that Zimbabwe will increasingly find itself with little food reserves to cover drought periods.

Box 3. Recommendation in the Zimbabwe Country Report

"There is a need for government to undertake a major review of land-use planning with due consideration given to an integrated resources management approach. Thus, the current exercise to assess Land Tenure Systems suitable for Zimbabwe should seriously consider the conferring of ownership to land owners together with the formal obligations on the part of the owner to use the land in a sustainable and productive way. Government agencies in charge of executing the resettlement program can also take into consideration the anticipated impacts of climate change" (Matarira et al., 1995).

5. Problem Analysis

5.1 Governance

The pre-eminence of political expediency over rational development goals by an entrenched political elite is not peculiar to Zimbabwe. It is almost the hallmark of many African political systems, which, for nearly half a century now, have committed national resources to internal conflicts driven by political agendas. Early in 2002, an incumbent president of a poor African country, with a per capita GDP of less than US$1,000 and an HDI of about three, blew up a vital highway bridge to maroon an electoral rival in a disputed poll. The decolonization of Africa left most of sub-Saharan Africa consisting of groupings of ethnic entities, some with strong traditional institutions, bound together by common borders. Such ethnic groupings were to play an important part in defining the nation state. In many countries, the emergence of the nation state took several decades and was characterized by internal feuds and armed conflict, a phenomenon still evident in many African countries. Forty-four years after the first decolonizing event (Ghana in 1957), these ethnicity problems still haunt Africa. In 1996, out of 53 nation states, 14 were involved in armed conflict; in 2000, the number had risen to 18 (author's estimates).

Under these circumstances, the preoccupation and skills of African leaders have been in the political arena, where development issues take a back seat. The continent has gone through periods when imported ideologies rather than pragmatic economic and social strategies were the instruments of governance. The consultative capacity of the state was weak, making it difficult for national consensual action on issues of common national interest, such as environmental management.

Although there are encouraging signs of transition from authoritarian rule to consensual democracy in a number of countries, the issue concerning climate change is the time horizon in which the economic and social environment would have stabilized sufficiently for climate change and other environmental issues to become mainstream public and state concerns. The projected trajectory of the evolution of appropriate governance in a number of scenarios is circa 2025 to 2050 (see, e.g., Achebe et al., 1990). Zimbabwe, for example, experienced nearly two decades of economic development. In the mid- to late 1980s, Zimbabwe was edging toward the economies-in-transition category, with a per capita GDP of US$2,300 (UNDP, 1997), in comparison, say, to that of China of about US$300 for the same period

(Ruoen and Kai, 1995). It had achieved an economic growth rate of circa 5 percent per annum. However, the entire fabric collapsed within a matter of months, because of the pre-eminence of political decisions over development goals.

Thus, there is a mismatch between the state of governance and the necessary institutional capacity to cope with the challenges posed by the climate change. Although climate change has been on the international agenda for nearly two decades now, and all African states are signatories to the relevant protocols, climate change is not necessarily a priority at the domestic policy level (Box 3); several other global environmental issues such as loss of biodiversity and desertification have yet to be indigenized.

5.2 Limited Resources

The inability of state governments to cope with disasters related to climate change and climate variability also stems from their limited resources and a general ill-preparedness to cope with or avert these disasters because of policy and relevant institutional inadequacies. The UNDP (1998) notes that though as many as 30 African countries have begun to show upturns in economic growth and progress toward democracy, the overall indices for poverty in Africa show a continent in which a large proportion live below the poverty line, lacking access to safe water, education and health care. Magadza (2001) noted that in the Nile basin states the mean doubling period for urban populations is 14 years, ranging between 4 years in Rwanda and 29 years in Egypt. Updating infrastructure and services for such growth rates is simply beyond the capacity of most African economies. Thus clean water supply, education, waste management, health services and energy strategies lag well behind population pressures.

In analyses of the poverty syndrome in Africa, many authors have often cited access to land as one factor in rural poverty (Malentlema, 1993; UNDP, 1998; FAO, 2000). Zimbabwe offers an opportunity for examining the impacts of different tenure systems. As noted above, investment and development in the communal lands have stagnated for over 75 years. The impact on land degradation is illustrated in Figure 3, a satellite image of Zimbabwe. The light areas are deforested communal lands, while the darker areas are commercial lands holdings or state land for nature conservancy. It is instructive to note the similarity in reflectivity between the Zimbabwe communal lands and the arid lands of eastern Botswana in the bottom left corner of the image. The figure shows, for whatever reason, that the land use

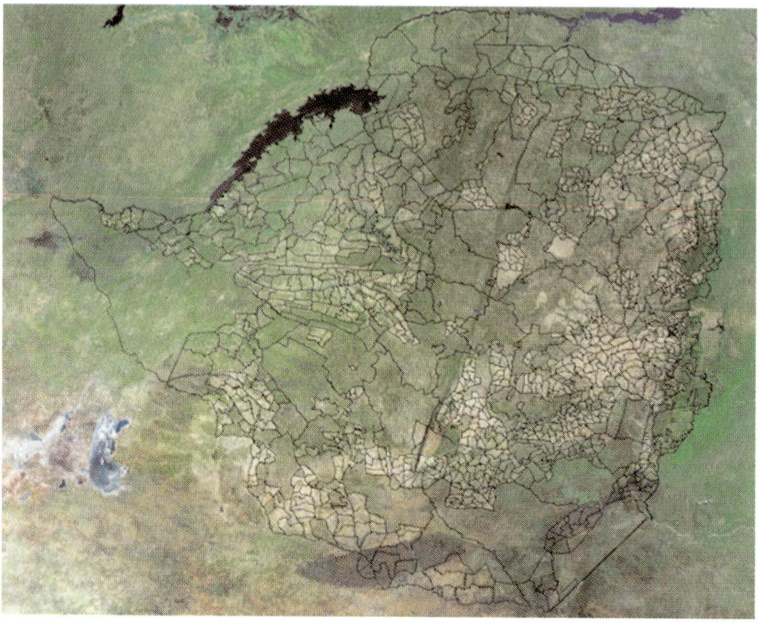

Figure 3. Satellite image of Zimbabwe. The light color areas are high reflective communal land areas. Note the similarity in reflectivity between the Zimbabwe communal areas and the arid lands of eastern Botswana in the bottom left corner of the image (Anonymous, 2002).

regime in the communal lands is not sustainable, and is leading to desertification. Decreased soil moisture, due to increased evaporative losses, will accelerate this process and greatly diminish Zimbabwe's capacity to feed itself, unless there is a major paradigm shift (subsistence agriculture) in land use. Half of the world's severely eroded land area (112 million hectares) is in Africa (Watson et al., 2000).

Figure 4, based on data from Muir (1997), compares agricultural productivity in communal tenure systems and the freehold tenure in Zimbabwe. While Zimbabwe has a historical political context that distorts land allocation, the figure nevertheless illustrates the disparity in productivity performance between nonmarket driven, usufruct land use tenure systems and market driven, entrepreneur titled tenure. The former is the most prevalent land use in Africa. In the commercial farming of Zimbabwe, nearly 90 percent of production is sold to the national food storage facilities as surplus production, in comparison to 41 percent from the

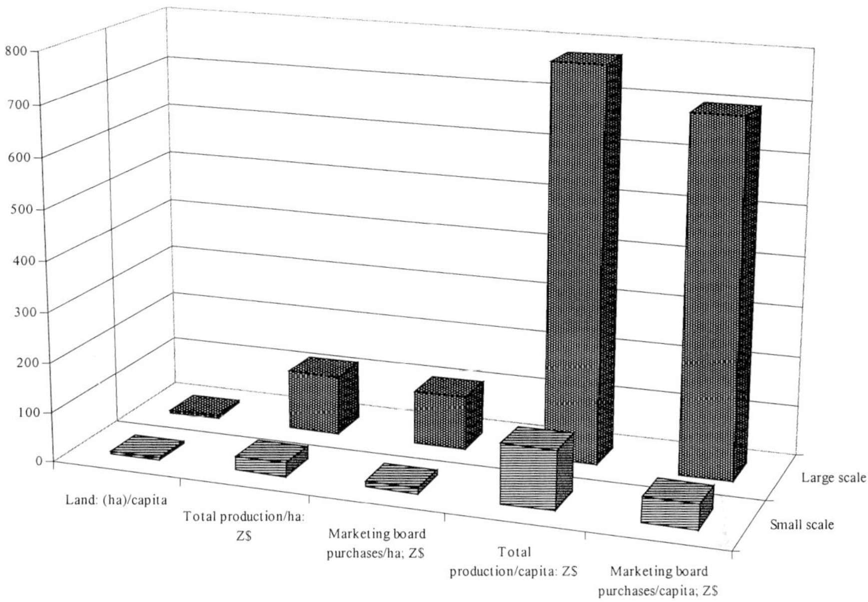

Figure 4. Comparison of small scale (communal land) and large-scale (commercial lands) production per unit of land and per capita in Zimbabwe. Data reconstructed from Muir (1997).

communal lands. Both production per unit of land and production per capita are higher in the commercial tradition (4.5 times higher for production per hectare and 6.6 times higher when reckoned per capita).

The second revelation from the data cited by Muir is that the average land holding for the communal lands is 4.66 hectares per person. Though the food crises in Africa are sometimes presented as issues related to land scarcity due to population pressure, few farmers in Japan have access to that much land, yet they live well. The poor performance of African agriculture is largely due to poor management, lack of technical and financial support and a diffuse marketplace.

The message from Muir's data is that African agriculture is currently underperforming and that climate change will impact on an already inefficient agricultural practice. Most land tenure systems in Africa reflect the status of land tenure at the end of colonial administration, and African governments have given little thought to the issue since then. In light of the recurrent food crises in African countries and the threats of climate change on agricultural systems, a review of land tenure with a view to ensuring security of investment by land users as well as providing technical services to farmers, is now required. For nearly a decade Zimbabwe communal farmers have voiced the need for security of tenure with legal rights. The IPCC prognosis on impacts of climate change on African food security is based on current practice and performance of the agricultural system; policy and technological innovations could greatly alter this prognosis.

6. Current Conditions

6.1 Technical Capacity

The TAR process revealed, at the time of the assessment, that there were sufficient data and, at the continental scale, critical capacity in Africa to assess climate change impacts. African states have prepared their country study reports in accordance with Article 4(a) of the UNFCCC in a timely fashion. Thus, it appears that Africa has the critical mass of human and information resources to appreciate the threats to sustainable development posed by climate change and to make policy formulations for climate change adaptations. The international community has made considerable investments for Africa's sustainable development in the sector of capacity building. The issue here is the failure of most African governments to capture this resource and engage it in planning for future sustainability.

6.2 Public Participation in Climate Change Dialogue

The hallmark of good governance is the capacity to engage the public in national discourse on issues affecting the nation. Few African governments have developed this capacity, relying mostly on their appointed officials, who do not necessarily, *ipso facto*, assume such responsibility. Consequently, the general African public is unaware of the climate change issue. I recently examined doctoral theses on ecological dynamics in Africa written by African authors, and they model ecological processes for as much as half a millennium into the future assuming no climate change.

6.3 Regional Institutions

Some climate change impacts such as hydrology and sea level rise occur at regional or river basin level. Magadza (2000) noted the need for regional institutions to address such impacts. Unlike in Western nations where institutions such as the European Community can be used as a common forum for policy resolution of regional and global environmental issues such as acid rain or global warming, or the Asia Pacific Seminar on Climate Change (Qureshi and Hobbie, 1993) and the Latin-American Energy Organization Casilla (Gutierrez, 1994), no facility at an equivalent stage of institutional development yet exists in Africa. Organizations such as the Southern African Development Community (SADC), the Economic Community of West African States (ECOWAS) and G77 may offer future promise. However, for such institutions to function effectively at both regional and intergenerational levels, political stability is essential. In the SADC region, for example, the regional responsibility for food security is entrusted to Zimbabwe; however, recent political developments in that country have undermined that responsibility.

7. Suggestions

7.1 Role of the Public

In the developed world, there is substantial public awareness of global climate change issues. State policies toward global warming are thus, to some extent, driven by public concerns and sectoral interests. In contrast, in Africa, there is very little penetration of awareness of climate change issues beyond state officials. There is thus little public demand for policy directives for climate change adaptation programs. Indeed, this lack of public awareness of climate change was identified as one possible thrust of the SysTem for Analysis, Research and Training (START) programs at a START sponsored workshop in Nairobi in February 2001.

For protocol and diplomatic etiquette, the IPCC consults states through the appropriate state channels, but there is also an urgent need for parallel activities to engage the public and private sector through nongovernmental organizations and educational systems. An effective public awareness campaign on climate change must be embedded in formal and informal education programs. An example is the country studies programs, in which states have responded to requests for reports, providing USAID or some other source funded the activity. Few of these reports are known, even by government officials, in the countries of their origin.

It is only through sustained sensitization of the public to climate change issues that a *sui generis* concern for climate change and adaptive capacity can emerge.

7.2 Collaborative Research

One of the information gaps evident in the African regional chapter in the TAR (Chapter 10) is the lack of systematic, long-term data to detect trends in ecosystem, health and socioeconomic impacts, even where such data could have been gathered. We suggest that the current regional institutions such as SADC and ECOWAS could have regional climate change desks to network the collection and archiving of data in these aspects of climate change impacts, and to ensure data availability to policy makers and scholars. This would enable a better informed assessment of climate change impacts and thus lead to more scientifically based adaptation measures.

While the current regional organizations can be the basis for policy formulation in regional climate adaptation, these must be supported technically by mirror images of regional technical institutions to undertake long-term research and monitoring for regional climate change impacts and adaptation. The current START initiative (Box 4) could be a basis for such institutions.

7.3 Technology Transfer and Entrepreneurship

A good deal of adaptation to climate change lies in appropriate technological innovations. For example, Africa receives sufficient solar radiation to meet most family needs for water heating, lighting and low voltage appliances. Yet even though the first world solar conference was held in an African country, there is very little evidence of Africa making use of its abundant solar energy resources. Dwellings and workplaces continue to be linked to the national grid to access often inadequate power supplies. (Recent blackouts in Kenya are an example.) Thus, the promotion of solar power technology would be an obvious adaptation against overdependence on hydroelectric power, whose reliability is and will be compromised by climate change.

Pachauri and Barathan (1995) proposed the use of economic instruments such as energy pricing as a tool to achieve CO_2 emission reductions in

> **Box 4.** Comments on African participation in climate research
>
> Several ongoing projects are addressing questions of global change in Africa. Virtually all of these projects are funded from sources outside of the continent where global change research receives more financial support. However, each of the projects listed is either targeted at building capacity to undertake global change research within the region or has developed strong partnerships with African scientists. A number of regional projects in Africa have been developed within the International Geosphere Biosphere Programme (IGBP), which provides a forum for international scientific exchange. There are a significant number of trained scientists in Africa who have a major contribution to make to national and international global change research programs. Those working in the African continent recognize the strength found in collaborative research and the necessity of improving scientific equity within Africa. The SysTem for Analysis Research and Training (START) program, which is targeted at building capacity to undertake global change research in developing countries, provides an important step in this direction.

developing countries. While this may result in energy savings in urbanized sectors of society, many rural populations in Africa depend on biomass and animal waste for heating and lighting, with the respiratory malaises this entails. A meaningful development in solar power would benefit both urban and rural communities.

The reasons for this inertia against using technology and entrepreneurial solutions are twofold. First, Pachauri and Breather note that one of the impediments in energy cost saving schemes is the pre-eminence of state-run bulk energy producers whose economic strategy is to maximize the market for their product. The other is that development agencies have been waiting for governments to take the initiative on solar energy programs. While the state should provide the enabling environment for the adaptation and innovations in technology transfer, the process should be driven by entrepreneurship and market forces.

What the African entrepreneur needs is exposure through trade partnerships to the opportunities for technological innovations pertinent to climate change. UNEP (2000, Chapter 3) notes that "private foreign investment, concentrated in a limited number of developing countries, was about US$250 000 million, compared to overseas development assistance (ODA) of less than US$50 000 million". These numbers demonstrate the overriding importance of the private sector in the world's economy and consequently in environmental issues. Africa needs to tap into these trade trends with respect to climate change technology. The Department of Energy in Zimbabwe has produced a number of possible candidates for household energy initiatives, but they lie on the shelves for lack of industrial uptake. There is opportunity for quantum leaps in technological transfer here, rather than the low-level, donor driven "appropriate technology" paradigm. The Climate Change Technology Initiative (www.climatetech.net/shtml) is a possible vehicle for accessing technology innovations for adaptation in Africa.

8. Conclusions

In general, African states have little concern for climate change impacts. This is a reflection of a failure to transit from activities driven by political agendas to those driven by development oriented agendas, as well as limitations imposed by available financial resources. There is thus a lack of connectivity between climate change adaptation needs and state priorities. Yet the limited data available indicate that impacts related to climate change and climate variability could constitute a significant financial drain on the African states.

Though the state has the obligation to provide an enabling environment for climate change adaptation activities through policy instruments, the effective entry point to capturing the adaptation agenda is through technology transfer and entrepreneurship.

Because climate impacts are mostly regional, there is a need for regional, state and public forums for climate change discourse. There is a dire need in Africa for climate change public awareness. One way of enhancing public appreciation of the import of climate change impacts is for public accounts to reflect the cost of climate impacts to the state coffers.

References

Achebe, C., Hyden, G., Magadza, C. H. D and Okeyo, A. P. eds. (1990), *Beyond Hunger in Africa*. Heinemann, Nairobi, Kenya.

Anonymous (2002), Origins of the Land Crisis in Zimbabwe. Available at www.geog.umd.edu/LGRSS/Projects/zimbabwe/origin1.html.

Benson, C. and Clay, E. (1998), The Impact of Drought on Sub-Saharan Economies. World Bank Technical Paper No. 401. The World Bank, Washington, DC.

Desanker P. and Magadza C. H. D. (2001), Africa. Chapter 10 in *Climate Change 2001: Impacts, Adaptation, and Vulnerability, Contribution of Working Group II to the Third Assessment Report of the Intergovernmental Panel on Climate Change*, McCarthy, J.J., Canziani, O.F., Leary, N.A., Dokken, D.J. and White, K.S. (eds.), Cambridge University Press, Cambridge, UK.

FAO (2000), Global Information and Early Warning System on Food and Agriculture. Country: Zimbabwe, Special Alert No. 307. Food and Agricultural Organization of the United Nations, Rome.

Gleick P. H. (1993), Water in the 21st century. Chapter 9 in *Water in Crisis: A Guide to the World's Fresh Water Resources*, Gleick, P. H., ed. Oxford University Press, Oxford, UK.

Gutierrez F. J. (1994), The Regional Response to the Greenhouse Issue: Latin America and the Caribbean. Presented at the International Conference on National Action to Mitigate Global Climate Change, 7-9 June 1994, Copenhagen. UNEP Collaborating Centre on Energy and Risø National Laboratory, Roskilde, Denmark.

IPCC (1996), *Climate Change 1995: Impacts, Adaptations, and Mitigation of Climate Change: Scientific-Technical Analyses. Contribution of Working Group II to the Second Assessment Report of the Intergovernmental Panel on Climate Change*, Watson, R. T., Zinyowera, M. C. and Moss, R. H., eds. Cambridge University Press, Cambridge, UK and New York, 1-18.

IPCC (1998), Africa. Chapter 2 in *The Regional Impacts of Climate Change: An Assessment of Vulnerability*. Intergovernmental Panel on Climate Change, Cambridge University Press, Cambridge, UK.

Jallow, B. P., Toure, S., Barrow, M. M. K. and Mathieu, A. A. (1999), Coastal zone of The Gambia and the Abidjan region in Cote d'Ivoire: sea level rise vulnerability, response strategies, and adaptation options. In *National Assessment Results of Climate Change: Impacts and Responses*, Mimura, N., ed. Oldendorf Luhe, Inter-Research, Germany, 129-136.

Magadza, C. H. D. (1996), Climate change: some likely multiple impacts in southern Africa. In *Climate Change and World Food Security,* Downing, T. E., ed. Springer-Verlag, Dordrecht, The Netherlands, 449-483.

Magadza, C. H. D. (2000), Climate change impacts and human settlements in Africa: Prospects for adaptation. *Environmental Monitoring* **61**, 193-205.

Magadza, C. H. D. (2001), Emerging issues in sustainable water resources management in Africa. In *Lakes and Reservoirs as International Water Systems:*

Towards World Lake Vision. Jansky, L., Nakayama, M. and Uitto, J. I., eds. United Nations University, Tokyo, 14-27.

Malentlema, T. N., Compiler (1993), IUNS (International Union of Nutritional Sciences) report. In. *The State of Food and Nutrition in Africa, 1970s-1980s-1990s: Nutritional Anthropology: Prospects and Perspectives*. Ulijasek, S. J. and Strickland, S. S., eds. Smith-Gordon, London.

Mapande, R. L. (1995), The proposed Matebeleland Zambezi Water Project. In *Water Resources Use in the Zambezi Basin*, Matiza, T., Crafter, S. and Dale P., eds. IUCN (International Union for the Conservation of Nature and Natural Resources), Gland, Switzerland, 13-24.

MARA/ARMA (1998), *Towards an Atlas of Malaria Risk: First Technical Report of the MARA/ARMA Collaboration*. MARA/ARMA, Durban, South Africa.

Marshall, J., Kushnir, Y., Battisti, D., Chang, P., Czaja, A., Dickson, R., McCartney, M., Saravanan, R. and Visbeck, M. (2001), Review of Atlantic climate. *International Journal of Climatology* **21**, 1863-1898.

Matarira, C. H., Makadho, J. M. and Mwamuka, F. C. (1995), Zimbabwe: climate change impacts on maize production and adaptive measures for the agricultural sector. In *Interim Report on Climate Change Country Studies*, Ramos-Mañé, C., ed. U.S. Country Studies Program, Washington, DC.

McCarthy, J.J., Canziani, O.F., Leary, N.A., Dokken, D.J. and White, K.S. (2001), *Climate Change 2001: Impacts, Adaptation, and Vulnerability. Contribution of Working Group II to the Third Assessment Report of the Intergovernmental Panel on Climate Change*. Cambridge University Press, Cambridge, UK.

Muchena, P. and Iglesias, A. (1995), Vulnerability of maize yields to climate change in different farming sectors in Zimbabwe. In *Climate Change and Agriculture: Analysis of Potential International Impacts. Proceedings of a Symposium Sponsored by the American Society of Agronomy in Minneapolis, MN, 4-5 November 1992*. ASA Special Publication No. 59, 229-239.

Muir, K. (1997), Agriculture in Zimbabwe. Presented at the Leadership for Environment and Development (LEAD) International Zimbabwe Session—Cohort 4, April 21 to May 3.

Nicholson, S. E. (1998), Fluctuations of Rift Valley Lakes Malawi and Chilwa during historical times: a synthesis of geological, archaeological and historical information. In *Environmental Change and Response in East African Lakes*, Lehman, J. T., ed. Kluwer Academic Publishers, Kluwer Press, Amsterdam, 207-232.

OCHA (2000), ZIMBABWE—Floods—March 2000. OCHA-Geneva Contributions Report (as reported by donors). UN Office for the Coordination of Humanitarian Affairs, Geneva.

O'Malley, C. M. (1992), The biology of *Anopheles quadrimaculatus* Say. In *Proceedings of the Seventy-Ninth Annual Meeting of the New Jersey Mosquito-Control Association, Inc.*, Burlington County Mosquito Extermination Commission, Mount Holly, NJ, 136-144.

Pachauri, R. K. and Barathan, S. (1995), Slowing global warming: mitigation strategy for the developing world. *International Conference on National Action to Mitigate Global Climate Change. 7-9 June 1994, Copenhagen, Denmark.* UNEP Collaborating Centre on Energy and Environment. Risø National Laboratory, Roskilde, Denmark.

Qureshi, A. and Hobbie, D. (1993), *Proceedings of the Second Asia Pacific Seminar on Climate Change.* Climate Institute, Washington, DC.

Ruoen, R. and Kai, C. (1995), *China's GDP in U.S. Dollars Based on Purchasing Power Parity* World Bank Paper series 1415. World Bank, Washington, DC.

UNDP (1997), *Human Development Report 1997.* United Nations Development Programme, Oxford University Press, New York, and Oxford, UK.

UNDP (1998), *Progress against Poverty in Africa.* United Nations Development Programme, New York. Available at www.undp.org

UNEP (2000), *GEO-2000: Global Environmental Outlook.* United Nations Environment Programme, Nairobi, Kenya.

UNESCO (1997), Kenya National Seminar on Sustainable Coastal Development through Integrated Planning and Management Focused on Mitigating the Impacts of Coastline Instability (Whitesands Hotel, Mombassa, 23-25 June 1997)., *UNESCO Nairobi Office Bulletin* **32**, 14-15.

United Nations (1999), *Contributions for Natural Disasters in 1999.* UN Office for the Coordination of Humanitarian Affairs, Geneva.

Urbiztondo, R. J. (1992), Modelling of Climate Change Impacts on the Upper Zambezi River Basin. M. Sc. Thesis, University of Colorado, Boulder, CO, USA.

Watson, R. T., Noble, I. R., Bolin, B., Ravindranath, N. H., Verardo, D. J. and Dokken, D. J. (2000), *Land Use, Land-use Change and Forestry.* Cambridge University Press, Cambridge, UK.

Wigley, T.M.L., Hulme, M., Brown, O., Crossley, M., Mew, M., Salmon, M and Jiang, T. (2000), SCENGEN: A Global and Regional Climate Generator. (CDROM). Climate Research Unit, University of East Anglia, Norwich NR4 7TJ, United Kingdom.

World Bank (1990), Towards the Development of an Environmental Action Plan for Nigeria. Report No. 9002-UNI. World Bank, Washington, DC.

13. Adaptation, Sustainable Development and Equity: The Case of Pakistan

Shaheen Rafi Khan
Sustainable Development Policy Institute
Sector G-5, Islamabad, Pakistan

Society's ability or inability to adapt successfully to climate change reflects its state of development. While the Third Assessment Report of the Intergovernmental Panel on Climate Change accepts this premise, it also appears to contradict itself by alluding to macro and micro strategies for adapting to climate change as if the two were disconnected. This paper, using Pakistan as an example, argues that the absence of sustainable human development creates social and ecological instability that will most likely be exacerbated by climate change. In fact, this paper posits that unsustainable development paths tend to block climate-specific adaptations. Case studies are used to illustrate the effects of policy neglect and intransigence on poor communities and complex ecosystems, both of which are highly vulnerable to the effects of climate change.

1. Introduction

The vulnerability of a given system or society to climate change is a function of both its physical exposure to climate change effects and its ability to adapt to these conditions (Chambers, 1989). The sensitivity of systems is also a factor. Similarly, Ausubel (1991), Raynor and Malone (1998) and Munasinghe (2000) state that the significance of climate variation or change depends on both the change itself and on the characteristics of the society exposed to it. Thus, vulnerability recognizes the role of socioeconomic (and ecological) systems in amplifying or moderating the impacts of climate change. The nature of these systems determines adaptive capacity and adaptability. Adaptive capacity refers to the ability to prepare for climate change in advance (as in anticipatory adaptation) and adaptability refers to the ability to respond or cope with its effects (as in reactive adaptation). Hazards may be a more appropriate term to use here. Changes in climate are either likely to evoke spontaneous responses or allow sufficient lead time for society to develop strategies to deal with them. This is much less likely to occur in the case of extreme events.

Society's ability or inability to adapt successfully reflects its state of development. Article 3.4 of the United Nations Framework Convention on Climate Change (UNFCCC) enjoins countries to promote sustainable development, so that they are prepared for and can deal effectively with climate change impacts. A more rigorous expression of this theme, articulated in the Third Assessment Report (TAR) of the Intergovernmental Panel on Climate Change (McCarthy et al., 2001), is represented by the conjunction of and synergies between development, sustainability and equity (DSE; Munasinghe, 2000). The TAR posits a strong coincidence between the imperatives of sustainable development and the requirements of enhanced adaptive capacity.

In other words, promoting sustainable development is tantamount to improving society's capacity to adapt to climate change. It also links the societal and the ecological aspects of vulnerability. Specifically, the absence of sustainable development can lead to environmental degradation and pollution, adverse health impacts and economic losses directly and, through the vehicle of poverty, a condition referred to as the poverty-environment nexus (Khan and Naqvi, 2000a).

This volume focuses on two adaptation strategies. The first strategy envisions enhancing macro-scale adaptive capacity, such as increasing wealth and improving education, income distribution, institutions and health care so as to improve countries' abilities to cope with climate change and other stresses. The second strategy is to enhance capacity to adapt at a micro or sector-specific scale, such as planning for sea level rise and developing heat or drought resistant crops, so as to enhance the ability to cope with specific impacts on climate-sensitive sectors. In both instances attention will be paid to improving the understanding of the determinants of adaptive capacity and their interactions in defining the ability of actors to adopt appropriate adaptation strategies at the micro-level and macro-level vulnerability to climate change, climate variability and abrupt climate change.

This echoes the approach of the TAR (McCarthy et al., 2001, p. 536), which states:

> Two general strategies on adaptation can be used. The first is macro strategy that involves rapid development. Sustainable and equitable development will increase income levels, education and technical skills and improve public food distribution, disaster preparedness and management, and health care systems in developing countries of Asia. All these changes could substantially enhance social capital and reduce the vulnerability of these countries to climate change.
>
> The second strategy is a micro strategy that involves modifying the management of sectors that are most sensitive to climate change. This approach entails developing new institutions or modifying existing institutions related to these sectors that promote rather than discourage adaptation to climate change. It also involves modifying climate-sensitive infrastructures that are already planned or implemented or other long-term decisions that are sensitive to climate to incorporate the risks of climate change.

The point of departure in this paper is that the first strategy is a precondition for the second, that there are risks inherent in adopting one without the other already in place. The links are illustrated with the help of two case studies. The first case study examines adaptive response capabilities (adaptability) to extreme events (droughts and floods). Unsustainable development processes leave vulnerable communities exposed and unable to cope. Poor institutional responses are another aspect of unsustainable development. This provides worrisome signals for the future, when both the incidence and severity of such events are expected to increase.

Large dams in Pakistan are another illustration of the failure of micro strategy when macro strategy is inappropriate: the planning of such large infrastructure projects has tended to ignore social and environmental norms. Further, the existing policy parameters support building more dams rather than modifying existing structures to accommodate climate change impacts, even though these serve the purpose of water retention equally well without the severe social and environmental consequences of dams.

Section 2 of this paper presents climate change scenarios for South Asia, as presented in the TAR, focusing on both climate change and extreme events. Section 3 assesses Pakistan's performance in the light of sustainable human development criteria. Section 4 presents the two case studies, illustrating how the absence of development, sustainability and equity has left institutions unprepared in terms of their ability to both prepare for climate change and cope with its effects. The consequences are a prelude to what might happen as extreme events increase in frequency and intensity.

2. Climate Change Scenarios for Arid and Semi-Arid Asia

This section recaps Chapter 11 of the TAR (McCarthy et al., 2001) on arid and semi-arid Asia. Recent climate change scenarios generated by general circulation models (GCMs) for arid and semi-arid Asia are presented in Table 1.

Although area-averaged annual mean precipitation is projected to increase, this average alone can be misleading because it conceals an expected decline in summer precipitation over the central parts of arid and semi-arid Asia. This includes almost the entire Indo-Pakistan subcontinent. Because rainfall over this region is already low, severe water stress conditions—which can lead to an expansion of deserts—are quite possible with higher surface air temperatures and depletion of soil moisture. The largest reductions (precipitation reduced to <1 mm per day, 60 percent decline in soil moisture) are simulated in the arid regions of northwest India and Pakistan.

Figure 1 shows the climate regions of Pakistan, and Figure 2 shows the temperature zones of the country. Drought disasters are more frequent in years following El Niño/Southern Oscillation (ENSO) events. Figure 3 shows areas of drought in 2002. At least half of the severe failures of the Indian summer monsoon since 1871 have occurred during El Niño years. With any advanced anomalous warming of the western equatorial Pacific Ocean, such as that observed during the 1998 El Niño, a higher frequency of intense extreme events (both droughts and floods) across all Asia is possible.

Table 1. Changes in mean annual temperature and rainfall for arid and semi-arid Asia in 2050 and 2080 from baseline period 1961-1990

	Area-averaged annual mean warming		Area-averaged annual Mean rainfall	
Decade	Without sulfate aerosols	With sulfate aerosols	Without sulfate aerosols	With sulfate aerosols
2050	3°	2.5°	7%	3%
2080	5°	4°	11%	7%

Source: Data from McCarthy et al., 2001.

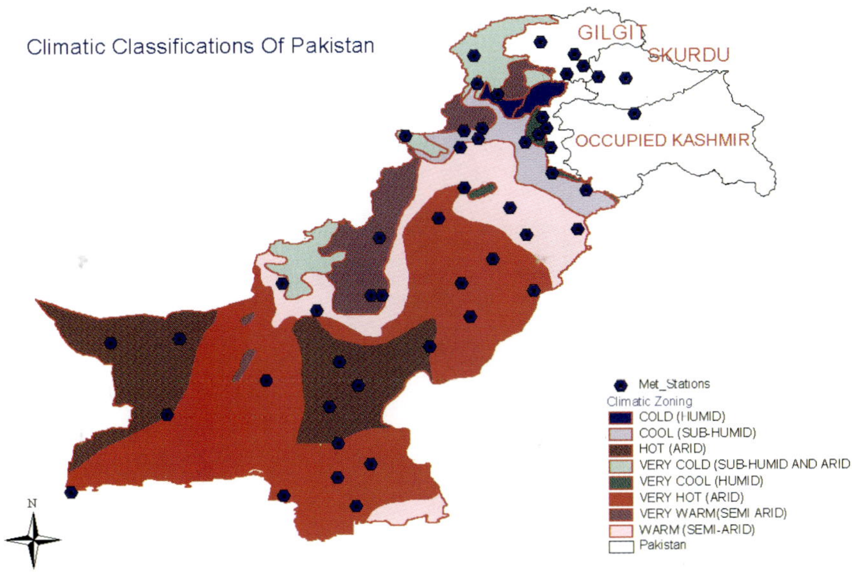

Figure 1. Climatic classifications of Pakistan

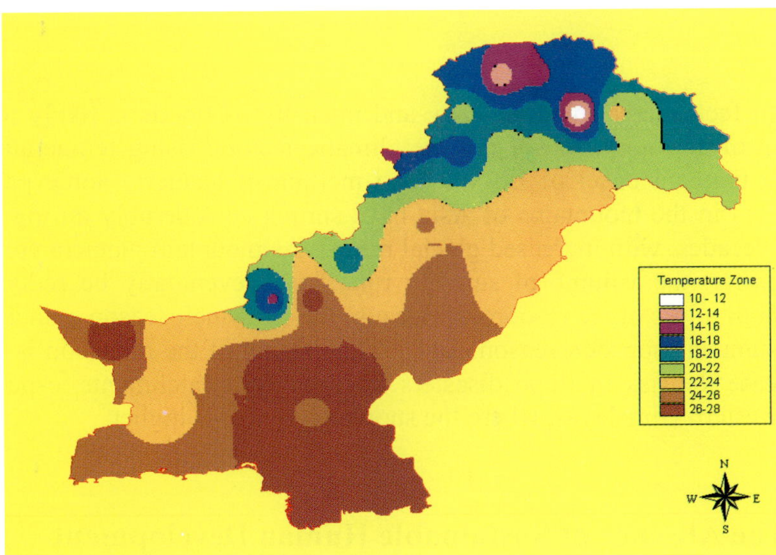

Figure 2. Temperature zones of Pakistan

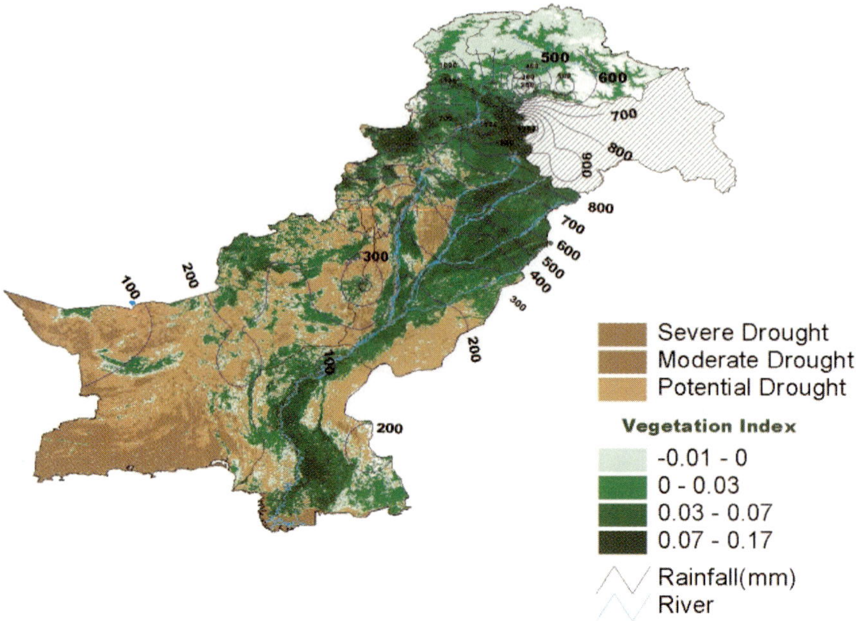

Figure 3. Areas of drought in Pakistan in 2002

In fact, increasing frequency and intensity of floods is likely to be the result of an interaction of diverse climatic factors: rising temperatures, the more frequent onset of El Niño and melting of glaciers. Some permanent glaciers in the mountains of Asia have shrunk considerably during the past few decades, with increased glacial runoff. As mountain glaciers continue to disappear, the volume of summer runoff will eventually be reduced as a result of a loss of ice resources. Because the snowmelt season coincides with the summer monsoon season, any intensification of the monsoon is likely to increase the risk of flood disasters in Himalayan catchments, especially in the western Himalayas, where the snowmelt runoff is higher.

3. The Absence of Sustainable Human Development

What are the consequences of neglecting sustainable human development? This inclusive term means development that is efficient, equitable and sustainable in terms, respectively, of resource use, resource access and resource resilience. "Resource" is used in its generic sense and refers to capital, human and environmental resources. For a more detailed discussion of the requirements and attributes of sustainable human development, see Khan (1998) and Munasinghe (2000).

The imperatives of sustainable human development are intergenerational justice, a concern with protecting the environment for future generations and intragenerational equity—ensured through social and economic empowerment which, among other things, is a prerequisite for sustainable livelihoods. This refers to the ability of communities to extract a living from the environment in a manner which does not degrade it, as well as to cope with the environmental excesses engendered by unsustainable development. Essentially a people-centered theme, it requires policy interventions to strengthen community coping and management capabilities.

Table 2 sheds some light on Pakistan's performance with respect to sustainable human development criteria. It illustrates how policy neglect, low financial allocations for conservation and nonconsultative and extractive management practices have weakened the coping capacity of communities and led to increasing environmental losses.

3.1 Impact on the Environment

Lapses from sustainable human development and the resulting deterioration of Pakistan's environmental health are evident from several key indicators. Roughly 38 percent of Pakistan's irrigated land is waterlogged and 14 percent is saline (Government of Pakistan, 2000a). Recent data indicate that more than 1 million acres of fertile, arable land in the Indus delta have become saline and unusable, largely due to the retention of freshwater flows by large dams across the Indus River (IUCN, 1997, and unpublished IUCN survey, 2000).The annual rate of deforestation ranges from 4 to 6 percent (Government of Pakistan, 2000b). Carbon dioxide emissions are increasing annually at a rate of 8-10 percent (Government of Pakistan, 1992). And an estimated 250 million gallons of untreated water from Karachi are dumped into the Arabian Sea every day, causing great harm to both humans and ecological systems. Water toxicity, triggered by the rural-urban interface, has begun to reach alarming proportions. While not fully documented, its debilitating health effects have begun to manifest themselves with increasing frequency, especially near large cities and towns.

Table 2. The absence of sustainable human development in Pakistan

Common elements	The three development strands		
	Economic	Social	Environmental
Development/ efficiency	Macroeconomic instability (fiscal deficits, current account imbalance, exchange rate overvaluation)	Low social sector allocations (education, health, water supply and sanitation, income generation and employment)	Policy neglect and low allocations for natural resource conservation and the use of environment-friendly technologies
	Sector price distortions (industrial protection via tariffs and licenses, input and output price subsidies, varying degrees of state intervention in industrial production, utilities service delivery, internal and external trade)	Poor outreach of poverty reduction schemes, politically influenced entitlements and misappropriation of funds	Failure to internalize environmental effects (perverse incentives encouraging extraction rather than conservation, unable to enforce/comply with environmental quality standards governing emissions and pollution)

Table 2. The absence of sustainable human development in Pakistan (cont.)

Common elements	The three development strands		
	Economic	Social	Environmental
Equity	Tax policies favoring the affluent (regressive indirect taxes, agriculture income tax exemptions)	Minimum provision for grassroots participation and empowerment of disadvantaged groups in the formulation and implementation of policies and development programs Centralized decision-making and governance provides little space for social capital to develop or articulate itself	Degradation and pollution impact disadvantaged groups adversely (ecological marginalization, urban industrial and vehicular pollution, household fuel consumption, sanitation and waste disposal)

Table 2. The absence of sustainable human development in Pakistan (cont.)

Common elements	The three development strands		
	Economic	Social	Environmental
Sustainability	Degradation of physical infrastructure (industrial, municipal, transport and utilities)	Lack of resilience in political systems (military coups frequently replacing elected governments)	Degradation—irreversible in many cases—and increased vulnerability (deforestation, rangeland degradation, biodiversity loss, coastal zone erosion, air, soil and water quality deterioration)
		Withering of informal legal and social systems thanks to modernization (panchayats, jirgas, traditional support networks and resource management practices)	

Source: Derived from Khan and Naqvi, 2000a.

Two attempts to quantify environmental losses raise serious concerns about the sustainability of economic growth in Pakistan. Using extremely conservative assumptions, World Bank economists Brandon and Hammond (Brandon, 1995) estimate that environmental degradation results in the loss of about 4 percent of GDP every year, thus offsetting considerably conventionally measured GDP gains. The breakdown of the GDP loss includes annual health impacts of water pollution (US$759 million) and air pollution (US$301 million), and the loss of agricultural production due to land degradation (US$300 million). Another way of interpreting the same data is in terms of the net savings rate of the country—after accounting for not only the depreciation of physical capital but also the degradation and depreciation of natural capital, including energy and mineral resource depletion, net deforestation and emission impacts. A low savings rate means the lack of adequate resources for investment and growth. The World Bank estimated the "genuine savings rate" for Pakistan in 1997 to be only 2.5 percent of GDP (World Bank, 1998).

3.2 Impact on Adaptive Capacity and Adaptability

Failure to achieve sustainable human development has left Pakistan vulnerable to the effects of climate change because of its low adaptive capacity and adaptability. This is also referred to as pre-adaptation vulnerability. The TAR's definition (McCarthy et al., 2001, p. 893) is "socially constructed or endogenous risks" which weaken the ability of society to deal with climate change or variability. Such vulnerability can be segregated by direct and indirect source and by sector. Efforts to address pre-adaptation vulnerability tend to strengthen society's ability to deal with climate change.

The Direct Source of Vulnerability: Institutional Failure

The existing state of degradation and pollution in Pakistan is a result of the institutional failure to mainstream environmental activities within the development process (Jalal, 1993). Specifically, this refers not to the management of the environment per se but to the management of development activities within the assimilative capacity of the environment. At some risk of generalization, one might define this process as highly resource, capital and technology intensive. Relatively little weight is given to economic and financial incentives as a means of conserving land, forest, air and water resources or using them in a sustainable manner. Modernization in and of itself contains impulses for degradation: for instance, when economic opportunity and mobility depreciate the value of the resource base for communities drawing their sustenance from it, or when national legal and regulatory systems override traditional, community-based resource management—in fact, are misused to exploit such resources unsustainably.

When such development takes place in the absence of democratic and decentralized governance, environmental problems are compounded. Centralized state institutions tend, at best, to be insensitive to the importance of community property rights and empowerment and, in general, to the need for public space to facilitate the growth of social capital. At worst, they are hostile to decentralized initiatives, viewing them as a threat to their sovereignty. This is unfortunate because, backed by the right incentives, traditional community practices offer considerable hope for sustainable environmental management (Jodha, 1998). Similarly, an aware and motivated civil society can play an active role in environmental conservation.

The Indirect Source of Vulnerability: Poverty

While Pakistan has achieved impressive aggregate economic growth—with a GDP annually averaging more than 5 percent over the past three decades and a steady increase in per capita income (currently assessed at a little over US$480 in real terms and at approximately US$2,000 in purchasing power parity)—the impact in terms of poverty reduction is more ambiguous. Poverty is a multidimensional term. Consumption poverty refers to the extent to which the private consumption of individuals or households falls below the "poverty line", the minimum acceptable standard of private consumption. Another important dimension focuses on human development, a term that captures improvements in education, health, water and sanitation and the provision of sustainable livelihoods. The proportion of the population below the poverty line fell from 46 percent in 1984-1985 to 34 percent in 1990-1991 (World Bank, 1995), but has increased to 40 percent in recent years. Less clear is whether the improvement in consumption poverty was accompanied by a reduction in the absolute number of people falling below the poverty line. Pakistan did not perform well by another measure of poverty, namely, income inequality as represented by the income share ratio and the Gini coefficient (Khan and

Aftab, 1997). Both have worsened over time. Finally, Pakistan's performance is even less creditable with regard to human development. Its social indicators are far below those of countries with comparable or lower levels of per capita income. Furthermore, disparities across provinces and across the rural-urban divide are pronounced (UNDP, 2000).

The socially constructed or endogenous risks identified by various sources fall into these two zones of pre-adaptation vulnerability (Table 3).

Vulnerability to Climate Change: A Sector Review

Table 4 presents areas or sectors in arid and semi-arid Asia most likely to be affected by climate change, as indicated in the TAR (McCarthy et al., 2001, Chapter 11), including an assessment of sector vulnerability. Clearly the existence of such vulnerability is likely to exacerbate climate change impacts. By the same token, caution is advised in attributing impacts to climate change which may, a priori, have been the result of unsustainable development processes.

Drawing on the regional level impacts, I assess the vulnerability of three sectors/areas:

- human settlements and human health
- ecosystems and biodiversity
- agriculture.

Each area/sector assessment looks at the consequences of institutional failure and low social sector priorities in Pakistan, illustrating how the response capabilities of poor communities to climate change and variability are thus weakened. In turn, by becoming environmental predators, the poor also become the instruments of destruction of their resource base, leaving it vulnerable to climate change impacts and ultimately weakening its ability to sustain them. It accelerates further the vicious cycle of the poverty-environment trap.

Human Settlements and Human Health. The key sources of human health problems in intermediate to large urban settlements are water pollution, air pollution and solid waste disposal. While water pollution caused by organic and chemical discharges is widespread, its impacts are predominantly on the poor, as a result of the skewed distribution of sewage, sanitation and piped water facilities in poor urban slums (World Bank, no date). Also, the poor are often forced to settle in low-lying urban areas which, although relatively more affordable, are more prone to flooding, water pollution and disease.

Table 3. Endogenous risks identified in the TAR

Indirect—poverty	
Improved access to resources	Ribot et al., 1996; Kelly and Adger, 1999; Kates, 2000
Reduction of poverty	Berke, 1995; Eele, 1996; Karim, 1996; Kates, 2000
Lowering of inequities in resources and wealth among groups	Berke, 1995; Torvanger, 1998
Improved education and information	Zhao, 1996
Diminished inter-generational inequities	Berke, 1995; Munasinghe, 2000
Improved infrastructure	Magalhães and Glantz, 1992; Ribot et al., 1996
Moderate long-standing structural inequities	Magadza, 2000
Direct—institutional	
Respect for accumulated local experience	Primo, 1996
Assurance that responses are comprehensive and integrative, not just technical	Ribot et al., 1996; Cohen et al., 1998; Rayner and Malone, 1998; Munasinghe and Swart, 2000
Active participation by concerned parties, especially to ensure that actions match local needs and resources	Berke, 1995; Ribot et al., 1996; Rayner and Malone, 1998; Ramakrishnan, 1999
Improved institutional capacity and efficiency	Handmer et al., 1999; Magadza, 2000

Source: Derived from Chapter 18 in McCarthy et al., 2001.

Air pollution ranks as high as water pollution in terms of its health effects, especially on the poor. Low-income neighborhoods mushroom around industrial areas and power plants, where exposure to air pollution is high. Traffic congestion and the resulting vehicular emissions are becoming an increasingly serious problem in the big cities. Poor communities are the most exposed to auto emissions and other toxic fumes, because they tend to live close to the main trunk roads (Government of Pakistan, 1992). Developing countries in particular are ill prepared to deal with the problem of solid waste. Once again, the burden of the problem falls disproportionately on the poor, a consequence of distorted municipal budget and planning priorities. Pakistan generates an average 50,000 metric tons of waste every day, and municipalities collect only 60 percent of this waste. In

Table 4. TAR projected impacts of climate change in arid and semi-arid Asia

Agriculture and livestock	Ecosystems and biodiversity	Desertification	Severe droughts and floods	Sea level rise	Human settlements and human health
Declining agricultural (food and fiber) productivity. Food insecurity. Large northward shifts of subtropical crop areas. Livestock: rangeland degradation reduces carrying capacity.	Ecosystem degradation: changes in the composition and distribution of vegetation types (grasslands, rangelands and woodlands). Exacerbating biodiversity loss.	Desertification: because of the current marginality of soil-water and nutrient reserves, some ecosystems in semi-arid regions may be among the first to show the effects of climate change.	Erosion and soil degradation	Sea level rise: Would cause large-scale inundation along the vast Asian coastline and recession of flat sandy beaches. Ecological security of mangroves and coral reefs around Asia would be put at risk.	Heat stress-related mortality and morbidity, principally in older age groups. Expansion of vector-borne diseases (malaria). Heightened vulnerability to skin cancers, cataracts and other ocular diseases. Increased vulnerability to waterborne diseases (cholera, diarrheal diseases).

Source: Derived from McCarthy et al., 2001.

general, the poorer city areas are the least served by garbage collection services, or not served at all. In fact, the solid waste collected in upper or middle class areas of cities is often dumped in the slums and city peripheries, either in landfills or directly into watercourses. The resulting problems are smell, disease vectors and pests attracted by garbage, and overflowing and clogged drainage channels.

Compounding the problem is the chronic lack of health facilities to deal with the effects of water and air pollution and unchecked waste disposal (both burning and dumping). The absence of adequate nutrition, the lack of education and overcrowded housing all increase vulnerability to disease. A telling statistic is that infant mortality continues to remain high, even though most demographic indicators have improved, because of exposure to waterborne diseases. The incidence of respiratory diseases and lead poisoning (predominantly among children) from mobile and stationary source emissions is escalating rapidly. Thus, those who are the most disadvantaged are ultimately made more so.

Ecosystems and Biodiversity. I focus on Pakistan's forests, especially in relation to developments in the northern temperate zones, where most of the country's primary forest stands are situated.

Between 4 and 5 percent of Pakistan's land mass is currently covered by forest. Despite this small coverage, deforestation is well under way. In particular, primary forest cover is declining, with associated losses in genetic diversity and resilience. Estimates in Pakistan's Biodiversity Action Plan (Government of Pakistan, 2000b) show woody biomass disappearing at between 4 and 6 percent per annum, which is feared to be the second highest in the world. Indeed, Pakistan is one of the developing countries with no remaining biologically undisturbed forests (WRI, 1997). Both the quality and the composition of forest stands are deteriorating. A recent study shows that good quality, tall tree forests (with more than 50 percent cover) occupy less than 400,000 hectares (Government of Pakistan, 2000b). Data collected through satellite imagery present an even a gloomier picture: only 308,000 hectares are under dense forest cover (NWFP, 2000).

The root cause of degradation lies in forest management practices that have focused more on economic utility than on environmental utility. Such practices are also detrimental to community subsistence needs. The British colonial government originally weakened community rights to the use of forest resources; usufruct rights remained but were heavily proscribed. Community management traditions, already fragile, have eroded further with new opportunities for employment and migration. Also, demographic and development pressures have forced communities out of their ancestral lands into marginal areas, where competition for resources is severe, resulting in further violations of indigenous property rights. The situation contains the seeds of conflict, with communities forced to act as predators rather than as guardians of the commons.

Weak property rights and increased fears of expropriation by powerful groups fuel insecurity about future earning streams, especially in an inflationary environment, and in general give rise to concerns about the free rider problem, namely, that forests will disappear no matter what the community might do (Inayatullah, 1996). Even in privately owned forests (*guzaras*) rights holders may see in regeneration a reintroduction of state property rights, which may stifle even natural regeneration (Azhar, 1993).

While there is little doubt that under the currently hostile management and tenure regimes communities are showing a propensity to raid forest resources, their activities pale in comparison with the activities of the so-called "timber mafia", commercial loggers willing to undertake illegal logging driven by rising timber prices. The timber trade also demonstrates a distinct anti-community bias; while communities are entitled to a substantial share of revenues (royalties) from the logging in the *guzaras*, active collusion between the "mafia" and the national forest department results in appropriation of the bulk of these royalties.

Deforestation drives home the point that if the root causes of degradation (deforestation in this case) are not addressed, even the beneficial effects of climate change will provide only a temporary reprieve. For instance, first-order impacts of increased atmospheric concentration, temperature and precipitation were evaluated with the help of BIOME 3 model simulations (Siddiqui et al., 1997). Of the nine biomes, three (alpine tundra, grassland/arid woodlands and deserts) showed reduction in their areas as a result of climate change and five (cold conifer/mixed woodland, cold conifer/mixed forests, temperate conifer/mixed forests, and steppe/arid shrublands) showed increases in their areas. There was no change in the area of xerophytic wood/scrub in the simulations. Enhanced CO_2 concentration in the atmosphere appeared to increase a biome's area, even in high temperature and low precipitation scenarios. However, when climate related gains are juxtaposed with socioeconomic pressures and resource mismanagement, such gains are likely to be undermined.

Agriculture. About 80 percent of the arable lands and 90 percent of agricultural output (which accounts for more than one-quarter of GDP) in Pakistan depend entirely on the Indus Basin Irrigation System (IBIS), which is the largest integrated irrigation network in the world. It consists of three major storage reservoirs, 19 barrages, 12 inter-river linking canals, 43 irrigation canal commands and more than 107,000 watercourses delivering water to the farms. Practically the entire system is unlined, resulting in system losses due to seepage, particularly in the watercourses. Annually, the system draws an average of 106 million acre feet (MAF) of surface water for irrigation, supplemented by another 45 MAF of pumped groundwater, much of which is by way of recovery from system losses due to seepage in areas underlain by fresh groundwater (Faruqee, 1999).

For the first 40 years or so of Pakistan's existence, agricultural policies were defined by the so-called "Green Revolution" strategy of agricultural growth. This strategy was premised on a sufficiency of both land and water. Major irrigation schemes (for example, the Mangla and Tarbela dams on the Indus) were launched to harness this water and to apply it at both the extensive and intensive margins (extensive margin means applying more water to new land brought under cultivation; intensive margin means an increase in cropping intensity—applying more water to the same piece of land). However, over the years the availability of land and water to be used in extensive cultivation has diminished progressively (Government of Pakistan, 1992). Against finite supply constraints, sectoral (agriculture, households, industry, energy) demand for water is growing. Pakistan is transiting rapidly from a water surplus to a water stressed country. A water affluent country in 1951, with an annual per capita availability of 5,300 cubic meters, Pakistan was expected to become a water scarce country by 2000, with an annual per capita availability of 1,300 cubic meters (Government of Pakistan, 1998). Table 5 shows that even under the most optimistic precipitation scenario, water deficits are likely to become large. These are projected to be even larger in the more probable declining precipitation scenario presented for the arid and semi-arid regions and in which Pakistan falls (Sustainable Development Policy Institute, 1997).

In terms of the impacts on agriculture, the primary concern is with food security. In the worst case scenario of increased warming and falling precipitation, the length of the growing season is reduced, especially in the arid areas, where crops are already on the margin of stress (Ahmed et al., 1997).

Table 5. Demand-supply water balances with climate change (MAF)

Year	2000	2010	2020	2050
Projected irrigation water demand	104.87	110.04	110.04	110.04
Households and industry	5.90	9.20	12.50	20.50
Total projected demand	110.77	119.24	122.54	130.54
Projected releases from dams	104.87	110.04	108.85	106.98
Projected deficit	5.90	9.20	13.69	23.56

Source: Derived from Ahmed et al., 1998.

Increased heat and reduced soil moisture also cause accelerated growth early in the season and affect the partitioning and quality of biomass. In combination, these factors reduce yield. This is offset by higher CO_2 levels, which have yield enhancing effects on C3 crops and to a much lesser extent on C4 crops like wheat, rice and maize. The net effects, therefore, are indeterminate. On the other hand, crops like rice and sugarcane would be hard hit by water scarcity. Spatial shifts in production are also predicted, especially in the crop zones dominated by wheat, rice, cotton and maize.

In the case of water, the two most critical—and related—interventions are water use efficiency and equity in its distribution. In the case of agriculture, it is adaptive research and improvement in extension services to poor farmers, because they cultivate areas that are already on the margin of stress and therefore are most vulnerable to adverse climate impacts. But the constraints imposed by the existing development biases are likely to hinder any such institutional and technical transitions.

Reforms of irrigation system management face formidable obstacles in the shape of institutional gridlock and feudal institutions, which have resisted efforts to increase water charges (*abiana*) and improve the quality of operations and maintenance (O&M). More critically, both have combined to subvert efforts at decentralizing irrigation system management and commercializing its distribution. Specifically, those efforts entail forming representative farmer organizations and transferring management of the tertiary irrigation and drainage system to them. The responsibilities would entail providing O&M of the distributories and minors from the main canals of the Indus Basin irrigation system, assessing and collecting water charges and distributing water among its members (see Khan et al., 2000b). The provincial irrigation departments are resisting the reforms because of perceived threats to jobs, perks and privileges. The large farmers see them as reducing opportunities for collusion with irrigation department officials in water theft and as undermining their traditional hold over small and landless farmers.

Agricultural research to increase adaptive capacity needs to focus on climate resistant varieties and environment friendly agronomic practices. Further, establishing a social-environmental interface with communities participating actively in environmental management is critical. However, the existing large and inflexible research infrastructure continues to remain locked into a "more of the same" mindset, namely, research into high-yielding, input-intensive hybrid varieties, which benefits large farmers primarily. With little research on the effects of current farming practices on soil and water quality, climate change driven research appears to be many years away.

4. Responding to Climate Variability: Case Studies

The case studies on droughts and floods demonstrate that weak adaptive responses are rooted in the neglect of the affected areas. An extension of such neglect is the failure to monitor the situation once floods and droughts are under way and the insufficient, uncoordinated responses to them. Emergency and mitigatory measures are carried out without involving the communities, creating new problems. Ultimately, response capabilities are a function of the level of preparedness, and this entails creating an infrastructure and service support network in the area. This would allow resident communities to benefit from opportunities which good years bring and have ready access to emergency relief and assistance during the disaster years. The case study of Kalabagh Dam illustrates that large infrastructure projects which disturb and fragment ecosystems increase the vulnerability of such systems to climate change impacts.

4.1 Adaptability: Droughts and Foods

Droughts

This section is based on an unpublished 2001 survey from SCOPE, a Pakistani nongovernmental organization specializing in desertification and other issues. The Thar, Cholistan and Khuzdar regions are arid areas, ecologically diverse and rich in biodiversity. Living conditions fall below subsistence levels. The inhabitants are primarily nomadic grazers, traveling vast distances with their livestock across sparse rangelands. Transient agriculture is practiced, with crops such as barley, maize, millet and vegetables grown as and when rains do come. There is little government infrastructure—schools, health centers, piped drinking water, roads. Women fetch brackish water for drinking from enormous distances, snake bite victims rarely survive because of lack of access to vaccines and food purchases from distant towns have to be planned carefully. A single rainless year drives residents far from their homelands. Ironically, the only time official interest is shown is during the hunting season, when functionaries help well-connected hunters take a heavy toll of the local biodiversity.

Arenji-Khuzdar: Arenji, a sub-*tehsil* of the Khuzdar district of Balochistan province, lies some 430 kilometers southwest of Quetta. About 35,000 people live in and around scattered villages in the sub-*tehsil*. Livestock raising and, more unpredictably, agriculture are the two sources of income. There have been no rains in the area since 1997. Prolonged drought has created a severe emergency situation affecting both people and livestock. Fatalities due to hunger, disease and malnutrition have been high.

Arenji illustrates the classic Sen (1981) syndrome of poverty amidst plenty. Compounding the effects of the lack of infrastructure and services, the government has made decisions that aggravate the consequences of its neglect of the area. An administrative order has banned interprovincial movement of wheat and flour, restricted individual purchases to 25 kg of wheat flour at one time and made it mandatory to produce identification at the time of purchase. Traditionally, one or a few individuals purchase necessities for the whole community, saving time and expense in traveling long distances. The new restrictions on both per capita entitlements and food movement have created scarcity. People are forced to eat degraded rice and wild vegetation. Instances have also been cited of starved buffaloes and sheep trying to eat dead animals. Malnutrition is on the rise and drastic drops in hemoglobin levels have been recorded, increasing vulnerability to disease. Contaminated water complicates the problem further.

Relief work is sporadic. Political party and religious activists, the Red Crescent Society and the government are all involved in relief efforts. However, the amount of relief is inadequate (the district administration has provided less than US$1,000 for the 35,000 affected people) and the work of the various relief agencies is not planned or coordinated.

Cholistan Desert: The Cholistan region starts from Bahawalpur, forming the southeastern part of the Seraiki Region in Punjab. Rohi, the Seraiki name of Cholistan, comprises immense deserts and plains renowned for their tranquility and harmony. Cholistan consists of two parts: Greater Cholistan and Lesser Cholistan. Greater Cholistan is a vast sand desert where human survival is extremely difficult. Lesser Cholistan, spread over a vast flat area, is relatively more populated, though here too the settlements are scattered. The word "permanence" has a different meaning in the Cholistani context. While never settling permanently in any specific place, searching constantly for water and pastures, the inhabitants cannot conceive of leaving their ancestral abode for the comforts of the city. However, over the last few years life has becoming increasingly difficult because of the drought.

The total population is estimated to be between 120,000 and 150,000 (the source documents vary in their estimates). The livestock population is roughly 11 times the human population. Mirroring the situation in Arenji, basic infrastructure, roads, public transport, electricity, health and educational facilities are nonexistent in Greater Cholistan. However, Lesser Cholistan does have some access to services because of its proximity to settled and developed areas.

Water is a vital resource given the extremely low rainfall in the area. The nomadic Rohians practice traditional rainwater conservation. Many reservoirs have been dug in the flat areas called *tobas* (an artificial lake in the Seraiki language). The *tobas* are identified by names and associated with clans which settle around them for as long as they last. There are also about 50 permanent settlements where deep wells have been dug. A good rainfall turns Cholistan green overnight, with an abundance of flowers and perennial grass. It also gives the people an opportunity to plant crops.

Clearly, infrastructure and services are vital to the area. In times of prosperity, they facilitate access to markets for livestock and related products. During lean periods, they are critical for human health, access to food and water and emergency relief assistance.

Thar Desert: The Thar desert is a vast dryland and home to 1.5 million people. The ecology and economy of Thar depend on rainfall. The grass and shrublands support large livestock populations and the inhabitants grow crops and vegetables in good years. However, a prolonged four-year drought has disturbed the desert's life cycle. In a break with pre-partition practice, the government has yet to declare the Thar desert a calamity area. Facing famine and disease, thousands of Thari families have started migrating to the Lower Indus Basin barrage areas in search of work and sustenance.

Floods

This section is based on Sustainable Development Policy Institute (2001). Extensive flood damage occurred along the Leh *nullah* passing through Rawalpindi because of heavy monsoon rains in August 2001. Although this flood was small compared to the scale of floods along the Indus River, the flood responses illustrate institutional inadequacies, especially since Rawalpindi is one of the major urban centers of Pakistan and situated close to the federal capital.

While the floods caused little loss to human life, 3,000 homes were destroyed, considerable assets and livelihoods were lost and diseases such as diarrhea and typhoid became endemic. The relief needs were, sequentially, meals and drinking water, dry rations and clothes, emergency health services, sanitation and clean-up and, finally, reconstruction and rehabilitation.

Yet the institutional response was minimal. Because of a lack of municipal oversight, people had constructed homes and shops in the low-lying areas, causing the *nullah* to become clogged with garbage and refuse. This proved to be a major factor in the damage wrought by the floods. Subsequently, the administration imposed a building ban across the board and offered to relocate the communities, who rejected the proposal outright. They do not believe the government will compensate them for the loss of their property, the investment of their lifetime savings. Second, relocation poses a major logistic and financial inconvenience (longer distances traveled to schools and jobs). By the same token, a survey should precede a building ban to determine where the unsafe areas are. Finally, technical solutions such as deepening the *nullah* reflect little awareness of social realities. The *nullah* is likely to become clogged again, unless communities are made part of a continuous sanitation program.

Voluntary donations of food water and essential supplies poured in, reducing the threat of starvation. However, ordinary citizens are not in a position to address the emerging health and rehabilitation problems. A number of nongovernmental organizations volunteered assistance in immediate relief operations, assessing the nature and extent of the damage, and rehabilitation.

The municipal government has limited resources, but people are willing to rebuild at their own cost. However, the government continues to adopt a somewhat hands-off attitude, which precludes constructive engagement with civil society. This is underscored by the fact that the majority of those affected are poor communities living in slum settlements and with little political clout.

4.2 Specific Adaptation: Kalabagh Dam

Large infrastructure projects continue to occupy center stage in the development schemes of most developing countries, and Pakistan is no exception. The Kalabagh Dam project is one such example, although it has been shown to be infeasible on both social and environmental grounds. Further, climate change reinforces the case against building the dam. However, institutional disregard of such considerations underscores the need to address generic development criteria rather than tinkering with projects and outcomes. Thus, the key issue is not whether climate change concerns can be integrated in such projects but, more fundamentally, whether the decision to build such projects is appropriate in the first place.

The Indus River, the agricultural lifeline of Pakistan, flows southwesterly for about 2,500 kilometers and empties through an immense delta into the Arabian Sea. The river system and its tributaries have provided Pakistan with some of the most fertile and best-irrigated land in the subcontinent, measuring about 200,000 square miles. Two large dams across the Indus River (Tarbela and Mangla) and a vast supporting irrigation network have both fragmented and altered its original ecosystems. A growing concern is that yet another dam across the Indus would prove environmentally disastrous, triggering irreversible degradation of an already fragile ecosystem.

The expectation is that Kalabagh Dam would store an additional 6 million acre feet (MAF) of water. Also, it would produce over 1200 MW of hydroelectric power. However, the global and regional context for assessing large dams like Kalabagh is changing, with conventionally described irrigation, flood control and energy benefits being viewed through the prism of sustainable development, and this provides the contextual backdrop for a review of the professed benefits (Khan, 1999).

Water Availability and Flood Control

In the first place, the water availability claims to justify a major dam at Kalabagh appear doubtful. WAPDA (Water and Power Development Authority), the government agency responsible for large water projects, used a wet cycle period to show higher inflows at the rim stations (146 MAF). The longer period, which includes both wet and dry cycles, indicates relatively lower inflows (139 MAF). A climate change scenario that includes increased temperature and reduced precipitation lends even less credence to surplus water availability claims.

There is also a perception in the minds of planners that large dams are the perfect flood prevention measure. The evidence for Pakistan shows otherwise; its large dams not withstanding, there has been no reduction in the incidence and intensity of floods or in the associated losses of lives, crops, livestock and infrastructure (Table 6).

Table 6. Flood damages in Pakistan

Year	Monetary losses (billion rupees in 1995 prices)	Number of lives lost	Number of villages affected	Area flooded (sq. miles)
1950	9.08	2,910	10,000	7,000
1955	7.04	679	6,945	8,000
1956	5.92	160	11,609	29,065
1973	5.52	474	9,719	16,200
1975	12.72	126	8,628	13,645
1976	64.84	425	18,390	32,000
1978	41.44	393	9,199	11,952
1981	N/A	82	2,071	N/A
1982	N/A	350	7,545	N/A
1988	15.96	508	100	4,400
1992	56.00	1,008	13,208	15,140
1995	7.00	591	6,852	6,518

Source: Derived from Ahmed et al., 1998.

There is no seeming pattern to the floods other than the fact that they could have coincided with wet cycles. In actual fact, the flood impacts appear to have been more severe after 1973, after the two major dams at Tarbela and Mangla were built. As Bayley (quoted in Abramovitz, 1996) has so aptly put it, "The 'flood pulse' is not a disturbance, flood prevention is". And that is exactly what large dams like Tarbela and Mangla are: disturbances on a large scale, which also supports the view that dams do not prevent floods, they merely create "flood threat transfer mechanisms" (Abramovitz, 1996). The solution is to work with communities, rely on their knowledge and supplement their flood mitigation and coping strategies. With a predicted increase in the frequency and severity of floods, owing to ENSO and melting glaciers, this logic is even more appealing.

Ecosystem Degradation

Degradation of the Indus delta ecosystem as a result of reduced water outflows is already a highly visible phenomenon. The present level of silt discharge, estimated at 100 million tons per year, is a fourfold reduction from the original level, before the rivers were dammed. The combination of salt-water intrusion (some reports show this reaches 30 kilometers inland, e.g., IUCN, 1997) and reduced silt and nutrient flows has changed the geomorphology of the delta considerably. The area of active growth of the delta has been reduced from an original estimate of 2,600 square kilometers (growing at 34 meters per year) to about 260 square kilometers. Freshwater reaches only a few of the creeks and others have become blocked. The delta is being transformed by strong wave erosion, an increasing dominance of sand at the delta front and an increase in wind-blown sand deposits as a result of losses in vegetation (IUCN, 1997).

The consequent ravages to the ecosystem have been exceptionally severe, in particular to the mangroves, which are its mainstay. They sustain its fisheries; act as natural barriers against sea and storm surges; keep bank erosion in check; are a source of fuelwood, timber, fodder and forest products; and are a refuge for wildlife and a potential source of tourism. Without mangroves and the nutrients they recycle and the protection they provide, other components of the ecosystem would not survive. In this case, the predicted climate change scenario is doubly harmful. It will not only reduce further freshwater outflows to the sea but also exacerbate the consequent salt-water intrusion via sea level rise.

4.3 The Crisis of Governance

The Kalabagh Dam typifies government insensitivity, in this particular case, to interprovincial concerns regarding water availability, environmental degradation and social displacement. An important factor in good governance is decentralized and consultative decision-making. By contrast, Kalabagh has been the very antithesis of this, with policy decisions being made in a highly centralized, often secretive, politically coercive and technically flawed manner. Indeed, WAPDA appears bent on an ex post vindication of a politically motivated step.

The present drought situation typifies this alarming blend of arrogance and ignorance. In defiance of all evidence against the dam, WAPDA continues to lobby for it, even as the Indus runs completely dry in the most improbable place, namely, at the confluence of the four main feeder rivers. Suggestions as outré as laser-melting the glaciers are gaining currency in the popular press. More substantively, the Sindh continues to have proportionate reductions in water, such that river flows downstream of Kotri to the sea have stopped completely (IUCN, 1997).

5. Conclusion

Society's ability or inability to adapt successfully reflects its state of development. The UNFCCC Article 3.4 enjoins countries to promote sustainable development so that they are prepared for and can deal effectively with climate change impacts. A more rigorous expression of this theme, articulated in the TAR, is represented by the conjunction of and synergies between development, sustainability and equity (Munasinghe, 2000). The TAR makes it clear that promoting sustainable development is tantamount to improving society's capacity to adapt to climate change.

The TAR proposes two general strategies on adaptation. The first is a macro strategy, which involves rapid development. Its elements are sustainable and equitable development aimed at increasing income levels, education and technical skills and improving public food distribution, disaster preparedness and management, and health care systems in developing countries of Asia. The second strategy is a micro strategy that involves modifying the management of sectors most sensitive to climate change. This approach entails developing new institutions or modifying existing institutions related to these sectors that promote rather than discourage adaptation to climate change. It also involves modifying climate-sensitive infrastructures that are already planned or implemented or other long-term decisions that are sensitive to climate to incorporate the risks of climate change.

In this paper I have attempted to show, with the help of two case studies, that the first strategy is a precondition for the second, that it is not possible to adopt one without the other already in place. The first case study involving the effects of drought and floods validates the TAR hypothesis that unsustainable development can leave poor vulnerable communities completely exposed and unable to cope. Poor institutional responses reflect another aspect of such development processes. This provides worrisome signals for the future, when both the incidence and severity of such events is expected to increase.

Large dams in Pakistan are another illustration of the failure of micro strategy when macro strategy is inappropriate. In the present development climate, the planning of such large infrastructure projects violates all social and environmental norms. Further, in their present configuration and numbers, large dams are likely to exacerbate adverse climate impacts rather than alleviate them.

References

Abramovitz, J. N. (1996), Imperiled Waters, Impoverished Future: The Decline of Freshwater Ecosystems. Worldwatch Paper 128. Worldwatch Institute, Washington, DC.

Ahmed, S. et al. (1997), *Climate Change Impacts and Adaptation Assessments in Pakistan: Agriculture Sector Study, Climate Change Impact Assessment and Adaptation Strategies*. Report Prepared for the UN Environment Programme, Islamabad, September.

Ausubel, J. H. (1991), A second look at the impacts of climate change. *American Scientist* **79,** 210-221.

Azhar, R. A. (1993), Commons, regulation and rent-seeking behavior: the dilemma of Pakistan's guzara forests. *Economic Development and Cultural Change* **42,** 115-129.

Berke, P. R. (1995), Natural-hazard reduction and sustainable development: a global assessment. *Journal of Planning Literature* **9(4),** 370-382.

Brandon, C. (1995), Valuing Environmental Costs in Pakistan: The Economy Wide Impact of Environmental Degradation. World Bank, Asia Technical Department, Washington, DC.

Cohen, S., Demeritt, D., Robinson, J. and Rothman, D. (1998), Climate change and sustainable development: towards dialogue. *Global Environmental Change* **8(4),** 341-371.

De Wet, N. (2001), *Adaptation*. International Global Change Institute, University of Wikato, Hamilton, New Zealand.

Eele, G. (1996), Policy lessons from communities under pressure. In *Climate Change and World Food Security*, Downing, T. E., ed. Springer-Verlag, Berlin, Germany, 611-624.

Faruqee, R. (1999), *Strategic Reforms for Agricultural Growth in Pakistan*. The World Bank, Washington, DC.

Government of Pakistan (1992), *The National Conservation Strategy*. Ministry of Environment—International Union for the Conservation of Nature, Karachi.

Government of Pakistan (1998), *Compendium of Environmental Statistics*. Federal Bureau of Statistics, Islamabad.

Government of Pakistan (2000a), *Agricultural Statistics of Pakistan*. Islamabad.

Government of Pakistan (2000b), *Biodiversity Action Plan for Pakistan: A Framework for Conserving Our Natural Wealth*. Worldwide Fund-International Union for the Conservation of Nature, Karachi.

Government of the NWFP (2000), *Provincial Forest Resource Inventory: Draft Final Report*. KFW-GAF-SIGNUM GmbH, Government of the North-West Frontier Province, Peshawar, Pakistan.

Handmer, J., Dovers, S. and Downing, T. E. (1999), Societal vulnerability to climate change and variability. *Mitigation and Adaptation Strategies for Global Change* **4,** 267-281.

Inayatullah, C. (1996), *Improvement of Governance in Forestry Sector*. Sustainable Development Policy Institute, Policy Brief, Islamabad, Pakistan.

IUCN (1997), *Large Dams: Learning from the Past, Looking at the Future, Workshop Proceedings*. International Union for the Conservation of Nature, Karachi, Pakistan.

Jalal, K. F. (1993), Sustainable Development, Poverty and Environment Nexus, Occasional Paper No. 7, Economics and Development Resource Center, Asian Development Bank, Manila, Philippines.

Jodha, N. S. (1998), *Poverty-Environmental Resource Degradation Links: Questioning the Basic Premises*. International Center for Integrated Mountain Development, Kathmandu, Nepal.

Karim, Z. (1996), Agricultural vulnerability and poverty alleviation in Bangladesh. In *Climate Change and World Food Security*, Downing, T. E., ed. Springer-Verlag, Berlin, Germany, 307-346.

Kates, R. W. (2000), Cautionary tales: adaptation and the global poor. *Climatic Change* **45**, 5-17.

Kelly, P. and Adger, W. N. (1999), Assessing Vulnerability to Climate Change and Facilitating Adaptation. Working Paper GEC 99-07. Centre for Social and Economic Research on the Global Environment, University of East Anglia, Norwich, United Kingdom.

Khan, S. R. (1998), Environment: A People's Perspective. Paper Prepared for the Third Sustainable Development Conference, Sustainable Development Policy Institute, Islamabad, Pakistan.

Khan, S. R. (1999), The Case Against Kalabagh Dam. Working Paper Series, Sustainable Development Policy Institute, Islamabad, Pakistan.

Khan, S. R. and Aftab, S. (1997), Structural Adjustment, Labor and the Poor in Pakistan. Working Paper Series, Sustainable Development Policy Institute, Islamabad, Pakistan.

Khan, S. R. and Naqvi, A. A. (2000a), The Environment-Poverty Nexus: An Institutional Analysis. Working Paper Series, Sustainable Development Policy Institute, Islamabad, Pakistan.

Khan, S. R. and Naqvi, A. A. (2000b), Regional and Social Environmental Assessment of the Proposed Punjab Irrigation System Improvement Project. World Bank-Sustainable Development Policy Institute, Islamabad, Pakistan.

Magadza, C. H. D. (2000), Climate change impacts and human settlements in Africa: prospects for adaptation. *Environmental Monitoring and Assessment* **61**, 193-205.

Magalhães, A. R. and Glantz, M. H. (1992), Socio-Economic Impacts of Climate Variations and Policy Responses in Brazil. United Nations Environment Programme, Esquel, Brasilia, Brazil.

McCarthy, J. J., Canziani, O. F., Leary, N. E., Dokken, D. J. and White, K. S., eds. (2001), *Climate Change 2001: Impacts, Adaptation and Vulnerability. Contribution of Working Group II to the Third Assessment Report of the Intergovernmental Panel on Climate Change.* Cambridge University Press, Cambridge, UK.

Munasinghe, M. (2000), Development, equity and sustainability (DES) in the context of climate change. In *Climate Change and Its Linkages with Development, Equity and Sustainability: Proceedings of the IPCC Expert Meeting held in Colombo, Sri Lanka, 27-29 April 1999.* Munasinghe, M. and Swart, R., eds. LIFE, Colombo, Sri Lanka; RIVM, Bilthoven, The Netherlands; and World Bank, Washington, DC, 13-66.

Munasinghe, M. and Swart, R., eds. (2000), *Climate Change and its Linkages with Development, Equity, and Sustainability. Proceedings of the IPCC Expert Meeting held in Columbo, Sri Lanka, 27-29 April 1999.* LIFE, Colombo, Sri Lanka; RIVM, Bilthoven, The Netherlands; and World Bank, Washington, DC.

Primo, L. H. (1996), Anticipated effects of climate change on commercial pelagic and artisanal coastal fisheries in the Federated States of Micronesia. In *Adapting to Climate Change: An International Perspective*, Smith, J. B., ed. Springer-Verlag, New York, 427-436.

Ramakrishnan, P. S. (1999), Lessons from the Earth Summit: protecting and managing biodiversity in the Tropics. In *Global Environmental Economics,* Dore, M. H. I. and Mount, T. D., eds. Blackwell Publishers, Oxford, UK, 240-264.

Rayner, S. and Malone, E. L., eds. (1998), *Human Choice and Climate Change, Volume 3: The Tools for Policy Analysis.* Batelle Press, Columbus, OH, USA.

Ribot, J. C., Najam, A. and Watson, G. (1996), Climate variation, vulnerability and sustainable development in the semi-arid tropics. In *Climate Variability, Climate Change and Social Vulnerability in the Semi-Arid Tropics,* Ribot, J. C., Magalhães, A. R. and Panagides, S. S., eds. Cambridge University Press, Cambridge, UK and New York, 13-54.

Sen, A. K. (1981), *Poverty and Famines.* Clarendon Press, Oxford, UK.

Siddiqui, K. M. et al. (1997), *Climate Change Impact Assessment and Adaptation Strategies for the Forestry Sector in Pakistan.* Pakistan Forest Institute, Peshawar.

Sustainable Development Policy Institute (1997), Baseline Socioeconomic Scenarios: Projections up to the Year 2020 and 2050. Climate Change Impact Assessment and Adaptation Strategies. Report Prepared for the United Nations Environment Programme, Islamabad.

Sustainable Development Policy Institute (2001), Flood Damage Assessment Report. Internal Brief Prepared by the SDPI Advocacy Section, Islamabad.

Torvanger, A. (1998), Burden sharing and adaptation beyond Kyoto: a more systematic approach essential for global climate success. *Environment and Development Economics* **3,** 406-409.

UNDP (2000), *Human Development Report.* Oxford University Press, New York.

World Bank (1995), Pakistan Poverty Assessment. Report No. 14397-PAK. World Bank, South Asia Region. Washington, DC.

World Bank (1998), *Pakistan Economic Update: Adjustment and Reforms for a Better Future.* World Bank, South Asia Region. Washington, DC.

World Bank (no date), A Review of Environmental Health Impacts in Developing Countries' Cities. Urban Management Program Discussion Paper Number 6. World Bank, Washington, DC.

WRI (1997), *World Resources.* World Resources Institute, New York.

Zhao, Z. C. (1996), Climate change and sustainable development in China's semi-arid regions. In *Climate Variability, Climate Change and Social Vulnerability in the Semi-arid Tropics,* Ribot, J. C., Magalhães, A. R. and Panagides, S. S., eds. Cambridge University Press, Cambridge, UK and New York, 92-108.

14. Enhancing the Capacity of Developing Countries to Adapt to Climate Change: A Policy Relevant Research Agenda

Richard J. T. Klein
Potsdam Institute for Climate Impact Research
Potsdam, Germany
Joel B. Smith
Stratus Consulting
Boulder, Colorado, USA

1. Introduction

Background

Global climate is changing because of human activities and will continue to do so. Even the most drastic measures to reduce the emissions of greenhouse gases will not prevent the further occurrence of impacts on nature and society.

Even before the inevitability of climate change was firmly established, the United Nations Framework Convention on Climate Change (UNFCCC) defined two main strategies to address climate change: mitigation and adaptation. Mitigation comprises all human activities aimed at limiting climate change by reducing the emissions or enhancing the sinks of greenhouse gases such as carbon dioxide, methane and nitrous oxide. Adaptation refers to any adjustment in natural or human systems that moderates harm or exploits beneficial opportunities associated with observed or expected impacts of climate change.

Until recently adaptation played only a minor part in the international climate change negotiations. Most national and international climate policies mainly focused on mitigation. However, the situation is changing. In its recent Third Assessment Report (TAR; McCarthy et al., 2001) the Intergovernmental Panel on Climate Change (IPCC) concludes that adaptation is now a necessary strategy in the face of climate change, along with mitigation. The report concludes that mitigation efforts will at best slow down climate change. In addition, the earth's climate is already changing, largely as a result of human activities.

The IPCC TAR also notes that the success of adaptation in averting dangerous impacts depends greatly on the adaptive capacity of the sectors and systems affected. Adaptive capacity is the ability to plan, prepare for, facilitate and implement adaptation measures.

The IPCC concluded that the most vulnerable countries and communities are those that are highly exposed to hazardous impacts of climate change and have limited adaptive capacity. Given the inevitability of impacts, enhancement of adaptive capacity is then a necessary condition for reducing vulnerability. The IPCC suggests that activities required for the enhancement of adaptive capacity are essentially equivalent to those promoting sustainable development.

The prominence of adaptation in policy discussions has been increasing internationally. Recent Conferences of the Parties (COPs) to the UNFCCC created several funds to support the building of capacity in developing countries and countries with economies in transition to address adaptation. COP-7 endorsed the option for the least developed countries to prepare national adaptation programs of action (NAPAs) to communicate urgent adaptation needs. NAPAs are supposed to communicate to donors the developing countries' priorities relating to adaptation. COP-7 recognized that "the rationale for developing NAPAs rests on the low adaptive capacity of LDCs" (UNFCCC, 2001). COP-8 gave more prominence to adaptation, noting in the "Delhi Declaration" noted that "…urgent action is required to advance adaptation measures" and providing specific guidance on implementation of the adaptation funds (UNFCCC, 2002).

The decisions of the COPs and the findings of the IPCC reveal the urgent need to strengthen research efforts aimed at a better understanding of adaptive capacity, how it relates to vulnerability to climate change and sustainable development, and how it should be enhanced. Without a clearer understanding of these issues countries run the risk of making inappropriate, inefficient, unproductive and even counter-productive adaptation investments that do not produce benefits as effectively or efficiently as intended.

As noted in Chapter 1, the workshop on "Enhancing the Capacity of Developing Countries to Adapt to Climate Change" was held to develop a research agenda to improve our understanding of adaptive capacity and how best to enhance it. The research agenda formulated by the workshop is synthesized in this chapter in six broad research areas. The research agenda reflects the importance of "mainstreaming" adaptation to climate change in sustainable development, which was widely acknowledged at the workshop.

This chapter reflects the consensus of discussions during the workshop. Where appropriate, we identify where participants had differing views on topics or where some views were expressed from a particular perspective.

2. Major Themes

The workshop identified a number of important themes that set the context for the research agenda:

- enabling environment, social capital and governance
- adaptive capacity
- natural hazards management and adaptation
- adaptation and developing country priorities.

These themes do not necessarily map into specific research topics, but are critical background for the topics.

Enabling Environment, Social Capital and Governance

The matter of "enabling environment" for adaptation is important for understanding the adaptive capacity of a society. The enabling environment includes all of the social structures that make it possible for a society to function, from the family unit to a government. An enabling environment is a prerequisite for adaptation to work effectively. Where there is a breakdown of civil order, such as in countries experiencing war, adaptation measures will be largely ineffective.

Social capital is "the norms and networks that enable people to act collectively." Social capital is different from economic capital in a number of ways. One is that it does not necessarily depreciate from time with use but, like a muscle, can be strengthened with use. A second is that more social capital is not necessarily desirable. Criminal gangs have a large amount of social capital, but it may undermine the social capital of others. Implicitly, there are optimal amounts and types of social capital.

A key factor in social capital is the role of the state, in particular, the role of the state in enhancing or impeding the development of social capital. There are interesting examples of communities developing social capital with or without support from the government. The examples range from the state playing a very small role to too large a role. In some cases social capital is developed at the community level, but many functions normally provided by a national government may not be taken care of. In other cases, the state sees social capital as a threat to its power and impedes its development. The latter case is an example of where adaptive capacity is diminished.

Such factors as level of development, social capital and specific adaptations are not separable. As noted above, specific adaptation measures will be unsuccessful if the enabling environment is inadequate. This could be for a number of reasons depending on the adaptation, such as specific projects being improperly designed, built or implemented or inadequately maintained. To be sure, simply having an appropriate enabling environment does not guarantee that it will be used effectively. Even an adequate enabling environment does not ensure that appropriate decisions will be made. So-called "maladaptations" may result.

Adaptive Capacity

It is the adaptive capacity of a society that essentially allows it to prepare for and cope with climate change. The main finding of the workshop was that specific actions such as protecting low-lying coastal areas against sea level rise or transferring technologies to enable developing countries to better adapt to climate change cannot be divorced from development or enhancement of an enabling environment for adaptation to climate change. In the absence of an enabling environment, specific adaptations will have limited effectiveness. Adaptive capacity is determined by a number of factors, including economic wealth, availability and access to technology, information and skills (information systems, training), institutions (e.g., inadequate institutions can limit adaptive capacity) and equity (e.g., access to resources) (Smit and Pilifosova, 2001). Possessing adaptive capacity does not ensure that it will be used well. As noted below, a key question is the degree to which investments should be balanced between enhancing adaptive capacity and making specific adaptations.

Natural Hazards Management and Adaptation

We can learn much from society's experience with coping with natural hazards. Indeed, the stimuli are not that different. The hazards research community has been addressing this topic for over 50 years. Recent work has shifted attention away from symptoms and structural solutions toward human causes and approaches that are more compatible with maintaining environmental services, that are sustainable and that address equity (Mileti, 1999).

As with natural hazards management, adaptation to climate change will be location and context specific. Although specific adaptation approaches will be useful in many different applications, the notion that "one size fits all" needs to be avoided. There are differences not only in vulnerability to climate change at the local scale but also in adaptive capacity. Differences in per capita income, distribution of income, governments, private sector, culture and other factors mean that adaptations need to be tailored to different situations.

Natural hazards management emphasizes adaptation at the community and regional levels because these are the scales at which adaptations are generally implemented. What is appropriate will vary from place to place. Rather than have one adaptation or set of adaptations for all regions and localities, it is important to determine what is most appropriate and works best within each location.

Adaptation and Developing Country Priorities

The case studies in this volume demonstrate that climate change is not always high on the policy agendas for most developing countries.[1] Their concerns are mainly, if not exclusively, focused on current problems such as reducing poverty and risk to current natural hazards. Given this situation, it is therefore unlikely that adaptation measures will be adopted if they are not integrated with ongoing policies and practices addressing other environmental and development issues, such as disaster management, combating desertification and conserving biodiversity. For example, in the Philippines, integrated coastal management, a tool that can incorporate long-term risks to coastal regions such as sea level rise and changes in cyclone intensity, is now being used to address current concerns such as illegal fishing and pollution. In addition, climate change can be integrated into central planning processes such as National Sustainable Development Plans.

Synergies can exist in developing an enabling environment for addressing developing country priorities along with reducing vulnerability to climate change. Other synergies may be created by strengthening monitoring and early warning systems, establishing sustainable and participatory resource use programs and implementing policies and measures aimed at reversing environmental degradation.

While developing countries generally do not see adaptation as a priority, some representatives of donor and multilateral organizations at the workshop expressed a need for developing better information to help the organizations set priorities. More specifically, they want to know which countries are most vulnerable to climate change and where adaptation can be the most effective. This is in the context of the UNFCCC and COP decisions that call for allocating funds on adaptation to the most vulnerable and least developed countries (two categories that are certainly not mutually exclusive). It was pointed out that where vulnerability is greatest may not necessarily be where adaptation will be most effective. Indeed adaptation may be most effective where adaptive capacity is greatest. This itself presents a conundrum with regard to directing resources, although the Convention language calls for supporting adaptation in the most vulnerable countries and regions.

Finally, the participants agreed that while it is important to address these research topics, it is also important that implementation of adaptation measures move ahead. There are many adaptations that have benefits under current climate and can be implemented effectively now.

3. The Research Agenda

Using the key themes discussed above as context, the following research topics are proposed. Research on these topics would help us better understand vulnerability to climate change and the process of adaptation so as to provide information necessary to assess the need for adaptation. In addition, the research topics address matters needed to help determine what adaptations can be most effective and how adaptation can be incorporated by developing countries. A substantial amount of research has already taken place on these topics. This list is not meant to imply that these are newly discovered questions, but that these are areas in need for further analysis so as to support the policy process. Of the six topics, the last three address policy directly, and how adaptation could be carried out.

How Does Adaptation to Climate Change Occur?

It is critical to understand the adaptation process, what forms adaptation to climate change can take, what processes and actors are involved and how it can be analyzed. Autonomous adaptation (responses that we expect affected individuals and sectors to take on their own in response to climate change) tells us what we can expect to happen in a business-as-usual manner, while proactive adaptation (actions taken in anticipation of the impacts of climate change) tells us how effective policy intervention can be.

With regard to autonomous adaptation, we need to improve our understanding of how effective it can be in reducing adverse effects of climate change and in taking advantage of positive effects. We also need to understand its limits. As mentioned above, this is needed to be able to determine vulnerability. One option may be to conduct more evaluations of adaptations to extreme weather and changes in climate variability to assess how they were addressed and the effectiveness, limitations and costs of interventions.

- *Who adapts to climate change and what motivates their adaptation?* There are a number of questions regarding adaptation. First, who adapts to climate change? Is it individuals, organizations, governments or the private sector? Second, what motivates their adaptation? Are they reacting to climate change itself or to intermediate effects such as change in crop growth or price? Do they perceive that climate is changing? At what levels is change in climate being perceived? Ministers, planners, system operators, farmers? At what level does change in climate, as opposed to good practice on adapting to current climate variability, need to be perceived?
- *What information do different actors need to aid or encourage them to adapt?* Information on changes in climate and consequences of such changes may need to be tailored to different actors. For example, farmers will require different information than water managers. There will be differences in the variables that are requested, e.g., peak precipitation, frost days and seasonal versus hourly information.
- *How can expectations of adaptive behavior be improved to account for differences in adaptive capacity?* Modeling studies of adaptation have tended to use simple assumptions about adaptive behavior, e.g.,

assuming little or limited adaptation or assuming perfect information, foresight and profit or social welfare maximizing behavior. A key research question is whether we can improve our understanding and ability to estimate and model adaptive behavior by individuals, private sector, organizations and governments. How will this differ from community to community? Can these models better account for the determinants of adaptive capacity? Can empirical studies be done, perhaps examining adaptation to climate variability or to long-term changes in other important factors (e.g., population growth) to improve our ability to predict future human behavior?

- *Who are the relevant stakeholders and what is their proper role in assessing adaptation options?* Stakeholders can play a central role in the assessment of adaptation options. Assessing adaptation options involves many values that stakeholders would be best at expressing. What methods are best to involve stakeholders in such analysis?

What is Vulnerability to Climate Change and How Should it Be Determined so as to Identify Adaptation Requirements?

After IPCC (McCarthy et al., 2001), vulnerability is a function of a system's exposure and sensitivity to climate change and its capacity to adapt to adverse impacts. Assessing the vulnerability of countries and communities to climate change is of policy relevance. Information gained in these assessments can be used to determine the need for adaptation. Therefore, it is of great importance to be able to define and assess vulnerability in such a way that informed and justifiable policy decisions can be made.

There is widespread knowledge about vulnerability across sectors and countries. This work has most recently been assessed by the IPCC TAR (McCarthy et al., 2001). Nonetheless there are still important gaps. In particular research on climate change vulnerability of developing countries has not been as extensive and rigorous as research on vulnerability of developed countries. While the various Country Study programs have supported vulnerability assessments in more than 120 developing countries and countries with economies in transition, workshop participants noted that vulnerability to climate change has not been universally assessed, particularly in Africa.

However, lack of coverage of vulnerability studies or even inadequacies in the studies that have been done should not serve as an excuse for not taking action. A fundamental message from the workshop is that much of what needs to be done to adapt to climate change was known well before climate change was considered. Many actions needed to reduce risk to current climate variability and to future climate change are already justified and can be implemented. Nonetheless it is important as these measures are taken to enhance our understanding of vulnerability so as to improve targeting of where further adaptation is most needed.

Determining vulnerability requires knowing *who* and *where* the vulnerable groups are, *what* they are vulnerable to and *when* and *why* they are vulnerable. In addition, it is important to understand how vulnerability may change over time as a result of development and other processes such as globalization. Developing an understanding of the causes and spatial and temporal characteristics of vulnerability (and thereby of the opportunities to reduce it) requires research that goes beyond the conventional vulnerability assessments carried out to date. This may require an understanding of the various economic, political, social and cultural factors that reduce or increase vulnerability and how these may change over time.

Most of the studies of vulnerability to climate change summarized in the First, Second, and Third Assessment Reports of the IPCC (Tegart et al., 1990; Watson et al., 1996; McCarthy et al., 2001) rely on model driven approaches to examining climate change impacts. They rely heavily on use of scenarios derived from general circulation models and application of quantitative models to estimate climate change impacts. As such, these studies tend to make somewhat simple assumptions about adaptation in response to climate change.

Whereas the first generation of vulnerability studies were characterized by model and scenario-based analyses of potential impacts, a second generation of vulnerability studies has emerged to complement the impact analyses with detailed assessments of adaptation and adaptive capacity. Second-generation vulnerability studies will build on the knowledge generated in the earlier studies to link a system's exposure to future climatic conditions with the system's vulnerability to today's climate. Much information can be derived from understanding how weather extremes affect countries, communities and sectors today and how they trigger adaptive responses, although some new extremes may be experienced. In addition, many of the impacts of climate change are not the result of extreme events, for example, shifts in areas at risk to spread of infectious disease or melting of glaciers resulting from long-term climate warming.

To enable the identification of realistic adaptation measures and options to reduce vulnerability to climate change, future studies should involve those most affected by climate change and those in a position to change practices and policies. Engagement with these stakeholders can provide useful insights into how current climate vulnerability is experienced and addressed, as well as information on opportunities and constraints to enhancing adaptive capacity. In addition, it places climate change in the broader context of environmental change and development, which is necessary to ensure that adaptation measures are compatible with policies addressing nonclimate issues.

There are a number of specific research questions under this broad topic:

- *What causes vulnerability to climate change?* It is critical to identify exactly what leads to vulnerability. If vulnerability is from current climate or climate change, what is the nature of the change? Is it change in mean conditions, frequency of extremes or intensity of extremes? Impact assessments to date have generally focused on changes in mean conditions and assumed no change in variance. While this captures some potential changes in extreme events, e.g., hot days are assumed to be hotter, it does not capture what may happen if frequency and intensity of extreme events change. Is vulnerability caused by inadequate development, inadequate enabling environment or specific maladaptations (e.g., developing in flood prone areas), or by a combination of these? Vulnerability is not just a function of climate but also a function of lack of adaptive capacity, interaction with other stresses and other causes.
- *Who is vulnerable to climate change*? Vulnerability to climate change can be quite different at different political, geographic, income, class and other scales. Vulnerability differs across countries and within countries as well as at regional, community, social group and household levels. Thus the extent of vulnerability may be quite different depending on the scale being analyzed. Assessments of vulnerability (and adaptation) have tended to focus on national or regional scale and (with a number of notable exceptions) not on smaller scales such as the village level. Vulnerability assessments need to address these differences in scale. In addition, assessments need to consider interrelated aspects of vulnerability. For example, the vulnerabilities of mangroves and communities dependent on fisheries should not be considered in isolation.

- *How does vulnerability differ depending on the time scale of climate change events?* The temporal scale of vulnerability also differs. Short duration extreme events present one set of vulnerabilities, while gradual changes that eventually exceed thresholds or carrying capacity can present another set of vulnerabilities. We can identify many vulnerabilities by examining the vulnerability of societies to current extreme events. Such vulnerabilities could be exacerbated by climate change. Beyond this, gradual or underlying climate change may present some new vulnerabilities and even surprises.
- *How does vulnerability change as societies develop?* Socioeconomic conditions in many developing countries could change quite considerably over coming decades as populations grow and economies develop. This in turn will change exposure and adaptive capacity and hence vulnerability. Understanding how different development pathways will affect adaptive capacity and vulnerability will improve our knowledge of vulnerability. This suggests conducting comparative studies to assess how different baseline conditions can affect vulnerability.
- *What is the proper role for stakeholders in assessing vulnerability?* As noted above, more recent assessments of vulnerability and adaptation have put more emphasis on involving stakeholders. This has been done for a number of reasons, but primarily to ensure that studies assess topics that are relevant and useful to those affected by climate change. Who are these stakeholders? How should they be involved in the analysis of vulnerability? What is their proper role in setting objectives and in carrying out assessments? In addition, attention must be given to stakeholders' motivations and incentives. Indeed, stakeholders could be provided capacity to conduct their own assessments.

A related set of questions address identifying relative vulnerability among or within countries so as to set priorities for adaptation. This was a controversial topic at the workshop and one that did not generate a consensus view. Nonetheless, some of the research topics relevant to this point are as follows:

- *What indicators could be developed or applied to identify need for adaptation and measure success of adaptation?* For mitigation, carbon is the common metric used to measure potential reductions in emissions and dollars per ton of carbon is the unit used to compare cost-effectiveness of mitigation options. What metric or metrics are appropriate for use in determining need and measuring success of adaptations?

- *What role can valuation play in indices of vulnerability or adaptation?* Expressing the value of natural resources in monetary terms or in terms of services provided can be helpful in understanding their importance. On the other hand, many of the systems affected by climate change do not exist in markets where monetary (or other equivalent) values are established. Many ecosystems that could be affected by climate change are not traded in the market (and even if they were the market value may not reflect their total value to society). In addition, subsistence and many other activities sensitive to climate change do not exist within markets and thus do not have market valuations associated with them. Another consideration is whether monetary values would be seen as legitimate by those participating in the process to set priorities for adaptation.

What Lessons Can Natural Hazard Management Provide for Adaptation to Climate Change?

Society's experience of coping with climate variability offers many valuable lessons and insights into how best to enhance capacity to adapt to climate change.[2] Indeed many aspects of climate change adaptation will address risk management, such as how best to prepare for and recover from extreme weather events. The experience of coping with such risks, particularly in the context of changing social conditions such as increased population, conversion of land to agriculture and urban uses, and changes in income and societal preferences, can offer many important insights into how to adapt to climate change. Of particular importance to research are the lessons we can learn about coping with changes in climate variability that have already happened.

- *How has the natural hazards community addressed changes in stress?* A simple perception of the difference between coping with current natural hazards and coping with climate change is that coping with natural hazards addresses static risks whereas coping with climate change addresses risks that are changing over time. This view is not correct. Natural hazards management must address changes in risk from changes in such factors as population growth, wealth, behavior and technology. In addition, extreme events change as a result of such forces as El Niño/Southern Oscillation (ENSO) or the Pacific Decadal Oscillation (PDO), which can alter the frequency and intensity of extreme events. How such changes have or have not been addressed can provide valuable lessons for how to adapt to climate change.

- *How has natural hazards management adapted to differences in circumstances around the world?* The workshop emphasized that adaptation is not a matter of one solution fitting all problems. Solutions need to be tailored to specific needs of communities at risk. There will be valuable lessons in examining how natural hazards management has been adapted to local circumstances. How much variation is there? What is the experience when common solutions are applied to different circumstances?

How Can the Enabling Environment for Adaptation Best Be Enhanced?

The importance of the enabling environment and such determinants as social capital is described above in the discussion on workshop themes.

- *How does enhancing the enabling environment contribute to enhancing adaptive capacity?* Adaptive capacity includes a well functioning enabling environment as well as wealth and access to technology and information. These are not mutually exclusive categories, and having an effective enabling environment may be a necessary but not sufficient condition for adaptation. An effective enabling environment can exist without such factors as wealth and access to technology and information. To what extent does enhancing the enabling environment contribute to enhancing adaptive capacity? If having an effective enabling environment is a necessary condition to be able to adapt to climate change, how much is necessary? More specifically, what attributes define an effective enabling environment? Are there declining benefits to even stronger enabling environments? Is having an effective enabling environment sufficient for being able to adapt to climate change?
- *What factors are important for enhancing the enabling environment?* As noted above, the enabling environment includes all of the social structures that make it possible for a society to function. Exactly what creates an effective or appropriate enabling environment needs to be clearly defined and done so in a manner that can allow for specific measures to enhance it.

- *What are appropriate roles for government, private sector and nongovernmental organizations in developing and enhancing an enabling environment?* To what degree should government, the private sector and nongovernmental organizations attempt to build or enhance social networks or allow local or regional networks to be established on their own? The degree to which individual or community initiative is encouraged or discouraged is key (i.e., weak versus strong government). It is also critical to look at the degree to which a state can enhance or impede development of social capital. A weak state may not provide necessary services to facilitate development of social capital and hence adaptive capacity. A totalitarian state can block development of social capital because it is seen as a threat, also reducing adaptive capacity, yet it can provide some services not provided in a freer state. What is the proper balance of state involvement?

How Can Adaptation to Climate Change Be Integrated into Sustainable Development and How Can Synergies with Other Policy Objectives Be Created?

Climate change does not exist in a vacuum. It is not as if we have solved all the world's problems and can focus our undivided attention on climate change. Indeed, in the midst of coping with global change, climate variability and other existing issues or problems, adapting to climate change in the developing (and developed) world is, as noted above, something that garners very little if any attention.

Sustainable development is a major guiding force. In 2002 the World Summit on Sustainable Development was held in Johannesburg, South Africa. At a previous meeting in Rio de Janeiro, Brazil in 1992, the United Nations Conference on Environment and Sustainable Development, the UNFCCC was signed.

The relevant research questions are as follows:

- *To what extent should sustainable development goals be modified to address climate change?* Will sustainable development in its current conception sufficiently enhance capacity to adapt to climate change? Do changes or modifications need to be made? For example, does sustainable development sufficiently account for such climate change impacts as inundation of low-lying coastal areas, changes in storm frequency and intensity, relocation of ecosystems and the need to change the character or location of economic systems such as agriculture, forestry or fisheries?

- *What sustainable development goals are justified even more by consideration of climate change?* Is the consideration of climate change an even stronger argument for certain sustainable development goals? Does consideration of climate change provide more justification for these goals (i.e., increase the benefits) or suggest that more resources should be invested in these goals?
- *How does equity affect sustainable development and adaptive capacity?* The issue of equity is an important consideration for both sustainable development and adaptation. Significant disparity in income distribution could be a barrier to effective adaptation. How distribution of resources and equity affects the capacity to adapt to climate change has received limited attention in the literature.

How Can Priorities for Adaptation to Climate Change Be Set?

Adaptation funding needs could well exceed available resources. At COP-7, the European Union, Canada, New Zealand, Iceland, Norway and Switzerland committed US$410 million for the adaptation funds created at COP-6bis.[3] With additional resources, it is expected that funding will be less than US$1 billion.[4] This is likely to be insufficient to meet all of the potential funding needs. Given limited resources, how should or can priorities for investing in adaptation be set?

With the recognition that available funding is unlikely to be adequate to meet all needs, there are somewhat different views about the breadth of what should be considered to be eligible for funding from the adaptation funds. One view is that adaptation to climate change should not be separated from other broader but quite related considerations such as encouraging sustainable development, reducing natural hazards and creating an enabling environment suitable to support implementation of adaptation measures. The basic argument is that since adaptation requires management of natural resources that are already subject to climate stresses, adaptation to climate change is not all that different from adaptation to current climate variability.

Another view is that the UNFCCC and Kyoto Protocol would not support merging of adaptation to climate change into these broad categories and that consideration of climate change adaptation could get lost if it was defined as, for example, being part of sustainable development. The merging of adaptation and sustainable development does not bring new directions to development, it just addresses the problems that are not being solved. This view asks for a clear and distinct set of priorities for adaptation to climate change. A question is whether adaptation presents a different set of issues than those addressed under sustainable development and other considerations such as natural hazards management. Does it introduce different stresses, i.e., new stresses or substantially different ones from those already being faced that justify some separating out of climate change adaptation from other activities?

There are a number of research questions relevant to the administration of the adaptation funds:

- *What criteria should be used to decide which projects are funded?* While this is very much a policy question, there are many aspects to this that are appropriate for research. For example, how can applicants demonstrate that proposed adaptations will be effective in reducing risks or adverse impacts they are intended to address? How do reviewers or funders of proposals compare the relative merits of proposals?
- *How can incremental funding be determined and assessed?* The issue of incremental funding, e.g., for GEF supported projects, is a controversial one. A number of participants expressed concerns with the concept. Yet it may be quite relevant for adaptation measures that overlap considerably with broader considerations such as sustainable development or natural hazards management. The question is then how to determine what is the proper contribution from the climate change community. This could address such matters as the degree to which the adaptation has been modified to account for climate change or the share of potential damages caused by climate change.

4. The Way Forward

The Potsdam workshop identified many important research issues regarding how best to enhance adaptive capacity in developing countries. Many of these research topics reveal our limited understanding of how societies will adapt to climate change if they are left on their own and of how international aid can best be channeled to most effectively enhance the capacity of developing countries to cope with climate change.

These research questions will take time to study. Yet the international community is likely to demand quick answers to these and other policy relevant questions about enhancing adaptive capacity. With the agreement at COP-7 in Marrakech, three new adaptation funds are being started and Annex I countries are being invited to make contributions to these funds. With recognition that the impacts of climate change are already being felt, there is little patience in the developing world for numerous studies on adaptation.

The challenge to the research community, to those funding research and to those conducting research is to address these research questions in a rigorous yet timely manner. Money on adaptation will be spent. The choice is to spend it with limited information on how best to enhance adaptive capacity or to inform the policy process on which investments are most appropriate.

Endnotes

[1] There are exceptions to this. For many small island states threatened with loss of water supplies or inundation of large parts of their land area by sea level rise, adaptation to climate change is a critical priority.

[2] Society's experience in coping with other climate related risks, such as those facing public health and ecosystem management, can also provide valuable lessons for adaptation to climate change.

[3] There are three funds that could be used to support adaptation in developing countries. The Adaptation Fund will come from the proceeds from emission reductions under the Clean Development Mechanism. The other two are the Special Climate Change Fund and the Least Developed Countries Fund.

[4] Presentation by Dr. Shuzo Nishioka at the Potsdam workshop, October 1, 2001.

References

McCarthy, J. J., Canziani, O. F., Leary, N. A., Dokken, D. J. and White, K. S. (2001), *Climate Change 2001: Impacts, Adaptation, and Vulnerability. Contribution of Working Group II to the Third Assessment Report of the*

Intergovernmental Panel on Climate Change. Cambridge University Press, Cambridge, UK.

Mileti, D. (1999), *Disasters by Design: A Reassessment of Natural Hazards in the United States*. Joseph Henry Press, Washington, DC.

Smit, B. and Pilifosova, O. (2001), Adaptation to climate change in the context of sustainable development and equity. In *Climate Change 2001: Impacts, Adaptation, and Vulnerability. Contribution of Working Group II to the Third Assessment Report of the Intergovernmental Panel on Climate Change*, McCarthy, J. J., Canziani, O. F., Leary, N. A., Dokken, D. J. and White, K. S., eds. Cambridge University Press, Cambridge, UK, 877-912.

Tegart, W. J. M., Sheldon, G. W. and Griffiths, D. C. (1990), Climate Change-The IPCC Impacts Assessment. WMO/UNEP Intergovernmental Panel on Climate Change. Australian Government Publishing Service, Canberra.

UNFCCC (2001), Conference of Parties 7. Decision 28/CP.7. FCCC/CP/2002/13/ Add.4. www.unfccc.int.

UNFCCC (2002), Decision_/CP.8: The Delhi Ministerial Declaration on Climate Change and Sustainable Development. www.unfccc.int/cop8/latest/ 1_cpl6rev1.pdf.

Watson, R. T., Zinyowera, M. C. and Moss R. H., eds. (1996), Climate Change 1995: The IPCC Second Assessment Report, Volume 2: Scientific-Technical Analyses of Impacts, Adaptations, and Mitigation of Climate Change. Cambridge University Press, Cambridge, UK.

Appendix—Workshop Agenda and List of Participants

Potsdam, Germany, 30 September-2 October 2001

Workshop Convenors
Potsdam Institute for Climate Impact Research, Potsdam, Germany
Stratus Consulting, Boulder, CO, USA

Workshop Sponsors
Canadian International Development Agency
Climate Change Unit, Environment Division, World Bank
Electric Power Research Institute, USA
Federal Ministry for Economic Co-operation and Development/
Gesellschaft für Technische Zusammenarbeit, Germany
United Nations Development Programme

Workshop Co-Sponsors
Adaptation and Impacts Research Group, Environment Canada
Netherlands Climate Change Studies Assistance Programme
United Nations Environment Programme

Workshop Rationale and Introduction
This workshop will develop an agenda for research on how best to enhance the capacity of developing countries to adapt to climate change (and related current climate stresses, although the emphasis will be on climate change). This research agenda should be relevant for governments and institutions that wish to support developing countries to adapt to climate change.

One approach to adaptation to climate change is to enhance "macro" scale adaptive capacity, such as increasing wealth and improving education, income distribution, institutions and health care so as to enhance countries' ability to cope with climate change and other stresses. A second approach is to improve the capacity to adapt at "micro" or sector-specific scale, such as planning for sea-level rise and developing heat or drought-resistant crops so as to enhance the ability to cope with specific impacts on climate-sensitive sectors. In both instances attention will be paid to improving the understanding of the determinants of adaptive capacity and their interactions in defining (*i*) the ability of actors to adopt appropriate adaptation strategies at the micro-level and (*ii*) macro-level vulnerability to climate change, climate variability and abrupt climate change.

A number of topics will be addressed at the workshop and in commissioned papers, including:

- Is vulnerability to climate change reduced more by investing in macro-scale adaptation or micro-scale adaptation?
- To what degree do differences in risks of climate change, non-climate stresses and levels of development lead to differences in adaptation requirements across countries and regions?
- How can adaptation assistance be tailored to be appropriate for local needs?

Workshop Venues

Sunday 30 September 2001	**Monday 1 and Tuesday 2 October 2001**
Hotel Steigenberger Maxx	Telegrafenberg, Building H, Room H1-H3
Allee nach Sanssouci 1	Wissenschaftspark Albert Einstein
14471 Potsdam	14473 Potsdam
Germany	Germany
Tel.: +49 (0)331 90910	Tel.: +49 (0) 331 2882702
Fax: +49 (0)331 9091909	Fax: +49 (0) 331 288xxxx

Workshop Agenda (27 September 2001)

Sunday 30 September 2001—Steigenberger Maxx Hotel

14.00 Registration of participants.

Session I: Adaptive Capacity: Analytical and Methodological Issues—Chair: John Schellnhuber

This session will set the scene for the workshop and for the research agenda that will be developed during and after the workshop. The presentations will briefly highlight important developments that have been made to date and provide an in-depth analysis of the various dimensions of adaptive capacity.

They will not necessarily report on particular research projects but draw together information from a range of sources to provide a balanced overview, leading to important outstanding research questions.

15.00 Opening and introduction—Richard Klein and Joel Smith.
15.30 From adaptation to adaptive capacity: findings from the IPCC TAR—Barry Smit and Olga Pilifosova.
16.00 Social aspects of adaptive capacity to climate change—Neil Adger.
17.00 North-South issues in adaptation to climate change: equity, sustainable development and technology transfer—Saleemul Huq.
18.00 Adjourn.
19.00 Workshop dinner with presentation by Ajay Mathur.

Monday 1 October 2001—Telegrafenberg, Haus H

Session II: Links and Integration with Non-Climate Issues—Chair: Holger Liptow

This session will provide the context in which adaptive capacity to climate change should be considered. Climate change is one of many issues that challenge sustainable development in developing countries. Enhancing adaptive capacity is likely to be most successful when strategies to do so are in line with other development policies and activities. The presentations in this session will highlight the various relevant non-climate aspects that need to be considered when analysing adaptive capacity to climate change.

11.30 The Global Environment Facility and its needs for scientific advice on adaptation—Shuzo Nishioka.
11.45 Adaptive capacity: what does it mean in the context of natural hazards?—John Handmer (by telephone).
12.30 Toward a vulnerability/adaptation science: lessons from famine early warning and food security—Tom Downing.
13.30 Lunch.
14.30 Economic aspects of adaptive capacity to climate change—Gary Yohe (by telephone) and Ferenc Tóth.
15.30 Where is the adaptive capacity to strengthen adaptive capacity?—Ian Burton.
16.30 Break.

Session III: Case Studies—Chair: Gretchen de Boer

The case studies in this session will provide real-life examples of studies, plans, strategies and experiences of enhancing adaptive capacity in developing countries. The case studies will highlight opportunities and barriers to enhance adaptive capacity, as well as related outstanding research questions. Other issues to be addressed include factors that determine a country's adaptive capacity, limitations to enhancing adaptive capacity and the links and integration between climate and non-climate issues. Case studies are from South Africa, India, the Philippines and Zimbabwe. Speakers have been asked to discuss extensively country-specific issues in their papers and focus their presentations on limitations to enhancing adaptive capacity, as well as knowledge gaps.

17.00 South Africa—Roland Schulze.
17.45 India—Anand Patwardhan.
18.30 Adjourn.

Tuesday 2 October 2001—Telegrafenberg, Haus H

Session III continued—Chair: Larry Williams

09.00 The Philippines—Rosa Pérez.
09.45 Zimbabwe—Chris Magadza.
10.30 Discussion on case studies.
11.00 Break.

Session IV: Towards a Common Research Agenda—Chair: Saleemul Huq

This session will evaluate the information and findings of the previous three sessions and address them in three interactive panel discussions, which will help to develop an agenda of research issues that are not currently being adequately addressed. The first panel discussion will focus on research opportunities whilst the second and third panel discussions will focus on research requirements from an NGO and institutional/donor perspective, respectively. The session will start with an overview of current research initiatives relevant to adaptive capacity.

11.30 Overview of current research initiatives—Richard Klein and Joel Smith.

11.45 Panel Discussion I: What are the important research questions?—Chair: Roger Street—Panellists: Daniel Bouille, Neil Leary, Isabelle Niang-Diop, Anand Patwardhan.

12.30 Lunch.

13.30 Panel Discussion II: What are the non-governmental interests?—Chair: Neil Adger—Panellists: Christoph Bals (Germanwatch), Madeleen Helmer (Netherlands Red Cross), Brett Orlando (IUCN), Larry Williams (EPRI).

14.15 Panel Discussion III: What are the institutional and donor interests?—Chair: Rosa Pérez—Panellists: Yasemin Biro (GEF), Kees Dorland (Netherlands), Arun Kashyap (UNDP), Philipp Knill (Germany), Liza Leclerc (Canada), Olga Pilifosova (UNFCCC), Avis Robinson (World Bank), Ravi Sharma (UNEP).

15.30 Break.

16.00 Discussion on research needs and opportunities.

17.00 Concluding remarks—Joel Smith and Richard Klein.

17.30 Workshop closure.

17.45 Press briefing.

19.00 Reception hosted by the Mayor of Potsdam in the Residence of the Marquis d'Argens, Neues Palais.

20.00 Concert of the Friderizianische Kammerorchester Potsdam in the Schloßtheater, Neues Palais.

List of Participants

Neil Adger
Tyndall Centre for Climate Change Research
University of East Anglia
Norwich, NR4 7TJ
United Kingdom
Ph.: +44-1603-59 37 32
Fax: +44-1603-50 77 19
Email: n.adger@uea.ac.uk

Christoph Bals
Germanwatch e.V.
Kaiserstr. 201
D-53113 Bonn
Germany
Ph.: +49-228-604 92 17
Fax: +49-228-604 92 17
Email: bals@germanwatch.org

Yasemin Biro
Global Environment Facility
The World Bank
1818 H Street, NW
Washington, DC 20433
USA
Ph.: +1-202-473 85 85
Fax: +1-202-522 32 40
Email: ybiro@worldbank.org

Gretchen de Boer
Scientific, Technical & Advisory Services Directorate
Canadian International Development Agency
200 Promenade du Portage
Hull, Quebec, K1A 0G4
Canada
Ph.: +1-819-994 17 80
Fax: +1-819-953 80 58
Email: gretchen_deboer@acdi-cida.gc.ca

Daniel Bouille
Bariloche Foundation
Piedras 482—piso 2° H
1070 Buenos Aires
Argentina
Ph.: +54-11-43 31 18 16
Fax: +54-11-43 34 47 17
Email:
danielbouille@arnet.com.ar

Nick Brooks
Tyndall Centre for Climate Change Research
University of East Anglia
Norwich, NR4 7TJ
United Kingdom
Ph.: +44-1603-59 39 04
Fax: +44-1603-59 39 01
Email: n.brooks@uea.ac.uk

Ian Burton
University of Toronto
72 Coolmine Road
Toronto, M6J 3E9
Canada
Ph.: +1-416-739 43 14
Fax: +1-416-739 42 97
Email: ian.burton@ec.gc.ca

Georg Caspary
Organisation for Economic Co-operation and Development
2 rue André Pascal
F-75775 Paris Cedex 16
France
Ph.: +33-1-45 24 96 18
Fax: +33-1-44 30 61 47
Email: georg.caspary@oecd.org

Cecilia Conde
(did not attend)
Centro de Ciencias de la Atmosfera, UNAM
Ciudad Universitaria,
Circuito Exterior 04510
Mexico, D.F. Mexico
Ph.: (52 55) 56 22 40 92
Fax: (52 55) 56 22 40 90;
56 16 07 89
Email: conde@servidor.unam.mx

Kees Dorland
Institute for Environmental Studies
Netherlands Climate Change Studies Assistance Programme
De Boelelaan 1115
NL-1081 HV Amsterdam
The Netherlands
Ph.: +31-20-444 95 55
Fax: +31-20-444 95 53
Email: kees.dorland@ivm.vu.nl

Thomas Downing
ECI
University of Oxford
1A Mansfield Road
Oxford, OXI 3SZ
United Kingdom
Ph.: +44-1865-28 11 80
Fax: +44-1865-28 11 81
Email: tom.downing@sei.se

Kristie Ebi
EPRI
3412 Hillview Avenue
Palo Alto, CA 94304
USA
Ph.: +1-650-855 27 35
Fax: +1-650-855 29 50
Email: krisebi@epri.com
and
WHO ECEH
Via Francesco Crispi, 10
I-00187 Roma
Italy
Ph.: +39-06-487 75 50
Fax: +39-06-487 75 99
Email: keb@who.it

Torsten Grothmann
Potsdam Institute for Climate Impact Research
P.O. Box 601203
D-14412 Potsdam
Germany
Ph.: +49-331-288 25 57
Fax: +49-331-288 26 42
Email: torsten.grothmann@pik-potsdam.de

John Handmer
(attended by phone)
RMIT University
Department of Geospatial Science
Building 12, Level 11
360 Swanston Street
Melbourne, Australia
Ph.: +61-3-9925 2213
Fax: +61-3-9663 2517
Email:
john.handmer@rmit.edu.au

Madeleen Helmer
Development Office, Climate Centre
Netherlands Red Cross
P.O. Box 28120
NL-2502 KC Den Haag
The Netherlands
Ph.: +31-70-445 57 03
Fax: +31-70-44 57 12
Email: mhelmer@redcross.nl

Saleemul Huq
International Institute for Environment and Development
3 Endsleigh Street
London, WC1 H0DD
United Kingdom
Ph.: +44-20-73 88 21 17
Fax: +44-20-73 88 28 26
Email: saleemul.huq@iied.org

Arun Kashyap
UNDP
304 East 45th Street
New York, NY 100017
USA
Ph.: +1-212-906 60 36
Fax: +1-212-906 51 48
Email: arun.kashyap@undp.org

Shaheen Rafi Khan
(did not attend)
Sustainable Development Policy Institute
No. 3, UN Boulevard
Diplomatic Enclave 1, G-5
Islamabad, Pakistan
Ph.: 00 92 51 22 78 134
Fax: 00 92 51 22 78 135
Email: shaheen@sdpi.org

Richard J.T. Klein
Potsdam Institute for Climate Impact Research
P.O. Box 601203
D-14412 Potsdam
Germany
Ph.: +49-331-288 26 51
Fax: +49-331-288 26 42
Email: richard.klein@pik-potsdam.de

Philipp Knill
Bundesministerium für wirtschaftliche Zusammenarbeit und Entwicklung
Friedrich-Ebert-Allee 40
D-53113 Bonn
Germany
Ph.: +49-228-535 37 49
Fax: +49-228-535 37 55
Email: knill@bmz.bund.de

Neil Leary
AIACC/START
2000 Florida Ave, NW, Suite 200
Washington, DC 20024
USA
Ph.: +1-202-462 22 13
Fax : +1-202-457 58 59
Email: nleary@agu.org

Liza Leclerc
Climate Change Environment Division
Canadian International Development Agency
200 Promenade du Portage
Hull, Quebec, K1A 0G4
Canada
Ph.: +1-819-994 39 24
Fax: +1-819-953 52 29
Email: liza_leclerc@acdi-cida.gc.ca

Bo Lim
National Communications Support Programme
UNDP-GEF
304 East 45th Street
New York, NY 10004
USA
Ph.: +1-212-906 57 30
Fax: +1-212-906 65 68
Email: bo.lim@undp.org

Holger Liptow
Deutsche Gesellschaft für Technische Zusammenarbeit (GTZ)
P.O. Box 5180
D-65726 Eschborn
Germany
Ph.: +49-6172-79 13 52
Fax: +49-6172-79 63 20
Email: holger.liptow@gtz.de

Chris Magadza
University of Zimbabwe
No. 6 Gatwick Close, Bluff Hill
P.O. Westgate
Harare
Zimbabwe
Ph.: +263-4-33 17 48
Fax: +263-4-33 17 48
Email:
profmagadza@utande.co.zw

Ajay Mathur
The World Bank
MSN MC 4-410
1818 H Street, NW
Washington, DC 20433
USA
Ph.: +1-202-473 25 61
Fax: +1-202-522 21 30
Email: amathur@worldbank.org

Elisabeth Mausolf
Deutsche Gesellschaft für
Technische Zusammenarbeit
(GTZ)
P.O. Box 5180
D-65726 Eschborn
Germany
Ph.: +49-6196-79 13 78
Fax: +49-6196-79 63 20
Email: elisabeth.mausolf@gtz.de

Isabelle Niang-Diop
Department of Geology, Faculty
of Sciences
University Cheikh Anta Diop of
Dakar
Dakar-Fann
Senegal
Ph.: +221-825 07 36
Fax: +221-824 63 18
Email: isabelle@enda.sn

Shuzo Nishioka
National Institute for
Environmental Studies
16-2 Onogawa, Tsukuba-Shi
305-8506 Ibaraki
Japan
Ph.: +81-298-50 23 31
Fax: +81-298-50 25 69
Email: snishiok@nies.go.jp

Brett Orlando
Biodiversity Programme
IUCN—The World Conservation
Union
Rue Mauverney 28
CH-1196 Gland
Switzerland
Ph.: +41-22-999 00 01
Fax: +41-22-999 00 25
Email: brett.orlando@iucn.org

Anand Patwardhan
Indian Institute of Technology Bombay
Shailesh J Mehta School of Management
IIT, Powai
Mumbai
India
Ph.: +91-22-576 77 88
Fax: +91-22-572 34 80
Email: anand@cc.iitb.ernet.in

Rosa Perez
Natural Disaster Reduction Branch
Philippine Atmospheric, Geophysical and Astronomical Services Administration
Agham Road
Quezon City 1100
Philippines
Ph.: +632-922 19 92
Fax: +632-922 19 92
Email: rose_perez_pagasa@hotmail.com

Olga Pilifosova
UNFCCC
SST Programme
Martin-Luther-King-Str. 8
D-53173 Bonn
Germany
Ph.: +49-228-815 14 28
Fax: +49-228-815 19 94
Email: opilifosova@unfccc.int

Lasse Ringius
UNEP Collaboration Center on Energy and Environment
Risoe National Laboratory Bldg. 142
Frederikborgvej 399
P.O. Box 49
DK-4000 Roskilde
Denmark
Ph.: +45-46 77 51 31
Fax: +45-46 32 19 99 9
Email: lasse.ringius@risoe.dk

Avis Robinson
Environment Department
The World Bank
MSN MC 4-410
1818 H Street, NW
Washington, DC 20433
USA
Ph.: +1-202-473 24 80
Fax: +1-202-577 21 30
Email: arobinson@worldbank.org

Hans-Joachim Schellnhuber
Potsdam Institute for Climate Impact Research
P.O. Box 601203
D-14412 Potsdam
Germany
Ph.: +49-331-288 25 01
Fax: +49-331-288 25 10
Email: schellnhuber@pik-potsdam.de

Roland Schulze
School of Bioresources Engineering
& Environmental Hydrology
University of Natal
Private Bag X01
3209 Scotville
South Africa
Ph.: +273-3-260 54 89
Fax: +273-3-260 58 18
Email: schulze@aqua.ccwr.ac.za

Ravi Sharma
United Nations Environment Programme
UNEP/DPDL
P.O. Box 30552
Nairobi
Kenya
Ph.: +254-2-62 42 15
Fax: +254-2-62 34 10
Email: ravi.sharma@unep.org

Barry Smit
Department of Geography
University of Guelph
Guelph, Ontario NIG 2WI
Canada
Ph.: +1-519-824 41 20
Fax: +1-519-837 29 40
Email: bsmit@uoguelph.ca

Joel Smith
Stratus Consulting Inc.
P.O. Box 4059
Boulder, C0 80306-4059
USA
Ph.: +1-303-381 82 18
Fax: +1-303-381 82 00
Email: jsmith@stratusconsulting.com

Susanne Stoll-Kleemann
Potsdam Institute for Climate Impact Research
P.O. Box 601203
D-14412 Potsdam
Germany
Ph.: +49-331-288 25 48
Fax: +49-331-288 26 20
Email: susanne.stoll@pik-potsdam.de

Roger B. Street
Adaptation & Impact Research Group
Meteorological Services of Canada
4905 Dufferin Street
Downsview, Ontario L6Z 3J7
Canada
Ph.: +1-416-739 42 71
Fax: +1-416-739 42 97
Email: roger.street@ec.gc.ca

Ferenc Tóth
Potsdam Institute for Climate Impact Research
P.O. Box 601203
D-14412 Potsdam
Germany
Ph.: +49-331-288 25 54
Fax: +49-331-288 25 70
Email: ferenc.toth@pik-potsdam.de

Anne de la Vega-Leinert
Potsdam Institute for Climate Impact Research
P.O. Box 601203
D-14412 Potsdam
Germany
Ph.: +49-331-288 26 29
Fax: +49-331-288 26 42
Email: anne.delavega-leinert@pik-potsdam.de

Roda Verheyen
Research Unit for Environmental Law
University of Hamburg—FB 02
Edmund-Siemens-Allee 1—Flügel West
D-20146 Hamburg
Germany
Ph.: +49-40-42 83 85 45 3
Fax: +49-40-42 83 82 63 5
Email: rodaverheyen@compuserve.com

Larry Williams
Global Climate Research
EPRI
3412 Hillview Avenue
Palo Alto, CA 94304
USA
Ph.: +1-650-855 26 95
Fax: +1-650-855 29 50
Email: ljwillia@epri.com

Gary Yohe
(attended by phone)
Wesleyan University
Economics Department
Middletown, CT 06459
USA
Ph.: +1-860-685-3658
Fax: +1-860-685-2781
Email: gyohe@wesleyan.edu